BLACK BODIES AND QUANTUM CATS

C014162461

Praise for *Black Bodies and Quantum Cats*

"To Jennifer Ouellette, physics is more human and intriguing and less arcane than people think, as she proves with her smart and varied collection of stories from the farther reaches of physics, written with a light touch."

SIDNEY PERKOWITZ, Candler Professor of Physics, Emory University, and author of *Empire of Light*

"I was so hooked by the end of the first chapter that I read it straight through. *Black Bodies and Quantum Cats* is a captivating look at how physicists think about the world. Even non-scientists may find themselves starting to think that way."

ROBERT L. PARK, Professor of Physics, University of Maryland

BLACK BODIES AND QUANTUM CATS

TALES OF PURE GENIUS AND MAD SCIENCE

Jennifer Ouellette

ONEWORLD

OXFORD

BLACK BODIES AND QUANTUM CATS

A Oneworld Book
First published in the USA by Penguin Books 2005
First published in Great Britain by Oneworld Publications 2006

Copyright © Jennifer Ouellette, 2006
Foreword copyright © Alan Chodos, 2006

All rights reserved
Copyright under Berne Convention
A CIP record for this title is available
from the British Library

ISBN-13: 978-1-85168-499-1
ISBN-10: 1-85168-499-9

Typeset by Jayvee, Trivandrum, India
Cover design by Two Associates
Printed and bound in Great Britain by Clays Ltd., St Ives plc

Oneworld Publications
185 Banbury Road
Oxford OX2 7AR
England
www.oneworld-publications.com

Portions of this book were published as "Pollock's Fractals"
(Discover, issue of November 2001), "Teaching Old Masters New Tricks"
(Discover, December 2001), and "Bubble, Bubble: The Physics of Foam"
(Discover, June 2002).

ILLUSTRATION CREDITS:

p. 11 Nicolaus Copernicus Museum, Frombork; p. 19 Illustration by David Wood; p. 24 The British Library; p. 31 Illustration by Patrick Moore; p. 41 LIGO Laboratory; p. 48 The Bakken Library and Museum; p. 53 Adventures in Cybersound/Russell Naughton; p. 62 The Bakken Library and Museum; p. 75 Daniel Bernoulli, *Hydrodynamica* (1738); p. 96 Foyle Special Collections Library, King's College, London; p. 99 Library of Congress; p. 115 Tesla Memorial Society of New York; p. 119 Adventures in Cybersound/Russell Naughton; p. 124 Museum of American Heritage; p. 131 Barry Truax, *Handbook for Acoustic Ecology*; p. 150 By permission of the Warden and Fellows, New College, Oxford; p. 151 Illustration by John Blamire, Brooklyn College/CUNY; pp. 161–2 Illustrations by Paul Dlugokencky; p. 185 Xerox Corporation; p. 200 Stanford University; p. 204 NASA/Johnson Space Center; p. 211 Museum of Science, Boston; p. 218 Laser Center, Hanover, Germany; p. 227 (top) CNF, Cornell University, (bottom) CCMR/CNF, Cornell University; p. 241 Photograph by C. D. Anderson, courtesy AIP Emilio Segrè Visual Archives; p. 255 The Library Company of Philadelphia; p. 284 Illustration by Paul Dlugokencky.

For Cami, Emily, Kathryn, Morgan, and Ryan,
may you always be curious about the world around you.

HAMPSHIRE COUNTY LIBRARIES	
CO14162461	
Bertrams	08.09.07
530	£8.99
068446	

Contents

Foreword by Alan Chodos, PhD   ix

Preface   xi

1   Renaissance Man   1
1509   Publication of *De divina proportione*

2   Talkin' 'bout a Revolution   9
1542   Publication of *De revolutionibus*

3   Lens Crafters   18
Circa 1590   Invention of the microscope

4   Good Heavens   26
January 7, 1610   Galileo discovers moons of Jupiter

5   Gravity's Guru   34
July 1687   Publication of Newton's *Principia*

6   Drawing Down the Fire   42
June 1752   Franklin's kite experiment

7   Trompe l'Oeil   50
1807   Invention of the camera lucida

8   Rules of Attraction   59
1820   First evidence of electromagnetism

9   Ingenious Engines   66
Circa 1840s   Babbage's analytical engine

10   Emperors of the Air   73
September 24, 1852   First dirigible takes flight

11 Disorderly Conduct   81
June 1871   Maxwell and his demon

12 Queens of Science   87
November 1872   Death of Mary Somerville

13 Calling Mister Watson   94
March 10, 1876   First documented transmission of human speech

14 Thrill Seekers   102
1884   First U.S. roller coaster opens at Coney Island

15 Current Affairs   110
May 1888   Tesla and electric power generation

16 Shadow Casters   118
February 2, 1893   Edison films a sneeze

17 Radio Days   127
1895   Tesla demonstrates wireless radio

18 Mysterious Rays   134
November 8, 1895   Röntgen discovers X-rays

19 A Thousand Points of Light   141
October 1897   Discovery of the electron

20 Quantum Leap   148
October 1900   Planck introduces quanta

21 It's All Relative   157
June 1905   Einstein introduces special relativity

22 Rocket Man   166
March 16, 1926   Launch of the first liquid-fuel rocket

23 That Darn Cat   173
1935   Schrödinger's quantum cat

24 Copy That   181
October 1938   First xerographic copy

25 Life During Wartime   188
July 1945   The Trinity test

26 Gimme Shelter   195
June 2–4, 1947   The Shelter Island conference

27 Tiny Bubbles   203
1948   Reddi-Wip appears on the market

28 Mimicking Mother Nature   210
May 13, 1958   Velcro trademark registered

29 Energize Me   217
December 1958   Invention of the laser

30 Small World   224
December 29, 1959   Feynman's classic Caltech lecture

31 Contemplating Chaos   231
Circa January 1961   Lorenz and the butterfly effect

32 Kamikaze Cosmos   238
July 1963   Discovery of cosmic microwave background radiation

33 When Particles Collide   245
April 1994   Discovery of the top quark

34 Checkmate   252
May 1997   Deep blue defeats Kasparov

35 Much Ado About Nothing   260
December 1998   Discovery of the accelerating universe

36 The Case of the Missing Neutrinos   266
February 2001   Solution to solar neutrino problem

37 Icarus Descending   272
September 2002   Schön found guilty of scientific misconduct

38 String Section   279
October 2003   Nova special on string theory

Bibliography   286
Acknowledgments   296
Index   298

Foreword

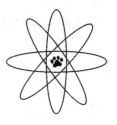

Physics has the reputation, not entirely undeserved, of being a difficult subject. Books about physics are often avoided, and that's too bad, because the questions that engage physicists are of interest to anyone with curiosity about the universe we live in.

The story of physics, as played out in the lives of physicists, inventors, and entrepreneurs, is deeply human and belies the stereotype of the scientist as guided only by strict rationality and objective reason. Progress in physics involves a lot of art as well as science. The way physics really happens cannot be found in the pages of professional journals alone.

In an astonishing performance, Jennifer Ouellette manages to weave all these elements together in a fascinating blend of history, science, and literary allusion. She takes the broadest possible view of physics, writing about milestones like the publication of Newton's *Principia*, but also about the invention of the roller coaster, and the quantum mysteries of Schrödinger's cat. As someone who was trained in English literature and has pursued a successful career as a science writer and editor, she brings an unusual set of talents to the creation of a unique work.

Jennifer and I have worked together for five years producing *APS News*, a publication that goes out monthly to professional physicists, the members of the American Physical Society. At the beginning of our collaboration, in the mood to try something new, Jennifer began a column titled "This Month in Physics History." I had thought that it might be devoted to events associated with the core physics curriculum that is usually taught in universities, but right from the start Jennifer was writing for a broader audience. Her first column, which survives in this book as the seed for Chapter 13, on the invention of the telephone, signaled her intention to move beyond the intellectual underpinnings of the subject and get out into the real world.

These columns have proved popular with the APS membership. We've received lots of letters adding to their content and challenging some of their assertions. We've also noticed, in traveling to physics departments around the country, that columns have been clipped from the paper and pinned to bulletin boards or people's doors. After about 50 columns had been written, we realized that, with a little reworking, they might find an appreciative audience among the general public.

Jennifer might not agree that only "a little reworking" was involved, but she has certainly written a book that is filled with compelling narrative and great science. Jennifer has mastered the difficult art of writing about science in a way that is both clear and correct, and beyond the science is her exceptional talent for telling a good story.

There are many good popular science books to choose from, but none quite like this one. It explores not only the high roads of physics history, but many of the more amusing byways as well. These essays are carefully and entertainingly written by an experienced science writer for a general audience. They are here to be browsed. Each one contains nuggets of information of both historical and scientific interest. We hope you enjoy them.

Alan Chodos, PhD

Associate Executive Officer
American Physical Society

Preface

How does a former English major and self-confessed "physics-phobe" end up writing an entire book about physics history? Frankly, I'm as surprised as you are. God may, or may not, play dice with the universe, but she clearly has a sense of humor. I spent the first 25 years of my life pointedly avoiding any and all exposure to the subject, even though I'd done very well in other science classes. Physics just seemed so alien and intimidating, something only a genius-level IQ would be able to fathom. Plus, there was all that math. Words = good; numbers = bad. That was my philosophy. If I hadn't stumbled into science writing quite by accident as a struggling freelance journalist on Manhattan's Lower East Side, I might never have been able to tell a quark from a cathode ray.

My story is not unique. Many of us, at one time or another, have bought into the notion of physics as an elitist, arcane high priesthood that only the most rarefied of intellects can ever hope to enter. But physics wasn't always set apart from the rest of society. It began as natural philosophy, and it was seen as part of a seamless whole, with the underlying mathematics woven into the very fabric of nature. The Renaissance visionary Leonardo da Vinci excelled in both art and science and would have found the current gaping chasm between those two spheres puzzling. To him, there was no distinction.

Somewhere along the line, we lost that notion of unity. Science and the humanities are often viewed as polar opposites, or as parallel lines (or universes) destined never to meet. Yet now, more than ever, physics is irrevocably shaping our notions of time, space, and our place in the universe. And this makes it relevant not just to specialists in the field, but to everyone.

The public is hungry for science: witness the astounding success of such bestselling books as Stephen Hawking's *A Brief History of Time* or Brian Greene's *The Elegant Universe*. Alas, more often than not, physics is presented in dry, jargon-filled prose that excises the crucial human element. Physics is a far cry from being a cold, hard discipline devoid of emotional content. Its history is replete not just with technological marvels and revolutionary ideas, but also with colorful personalities and human drama. Physicists are people, just like the rest of us, with all too human failings: they struggle, suffer, make mistakes and bad decisions, quarrel over petty issues, and sometimes even lie, or betray their colleagues. But they also persevere against all odds, driven by a touching idealism, and by a deep love for their subject that is nothing short of inspiring. As a teenager, I concluded that because I didn't have a gift for mathematics, the world of physics was com pletely closed to me. If only I'd known that the eminent nineteenth-century British physicist Michael Faraday also struggled with math; in fact, he was mathematically illiterate. That didn't stop him from becoming a brilliant experimentalist who pioneered the field of electromagnetism.

The job of a science writer is largely one of artful translation. Inevitably, something gets lost in the process. Richard Feynman, the "great explainer" of physics, once complained, "Many 'popular' expositions of science achieve apparent simplicity only by describing something different, something considerably distorted from what they claim to be describing." He is absolutely right. I have taken great pains to be accurate as well as engaging, and the science described in the chapters that follow is "correct" as far as it goes. But entire books have been written about each of the subjects at hand. It just isn't possible to explore all the branching complexities of a highly technical topic in 3000 words or less.

My aim in this book is not to write about physics as something separate from the broader culture, but to place it firmly within that context. So I use history, art, music, literature, theater, film, and television to help illustrate fundamental concepts, drawing on everything from the poetry of John Donne and Gerard Manley Hopkins to *The Da Vinci Code* and *The X-Files*. Physics is all around us, all the time. I see Newton's laws in the film *Addams Family Values*, and *Back to the Future* reminds me of the paradoxes inherent in special relativity. I find useful parallels between Edgar Allan Poe's "The Purloined Letter" and the mysterious nature of neutrinos. Jeanette Winterson's novel *Gut Symmetries* provides an elegant framework in which to sketch the outlines of modern string theory. Chapters are pegged to one particular idea or discovery, and arranged in chronological order to give the reader a sense of the developmental arc of physics throughout history.

It is my hope that both scientists and nonscientists will find something to savor herein. And perhaps there are others like my former self out there, whose eyes glaze over at the first sight of an equation and who hence assume that physics is inaccessible to them. To those readers, I say this: you don't need to have a PhD in physics or be a card-carrying member of Mensa to grasp the basic concepts presented in these pages—you just need a little patience and concentration to work through some of the trickier bits. You don't even have to read the chapters in order, although the later chapters will prove more accessible if you've read at least some of the earlier ones. I also hope that something will whet your appetite to learn more—in which case, I encourage you to peruse the bibliography. Question. Explore. Let your curiosity overcome your fear. You may be surprised to find (as I did) that the world of physics isn't such a scary, alien place after all.

1

Renaissance Man

1509
Publication of *De divina proportione*

Conspiracy and code breaking are the order of the day in Dan Brown's best-selling thriller *The Da Vinci Code*. In modern-day Paris, the curator of the Louvre Museum is brutally murdered, his body laid out at the foot of the museum's most celebrated acquisition, Leonardo da Vinci's *Mona Lisa*, in a pose reminiscent of Leonardo's equally famous sketch, *Vetruvian Man*. The crime scene is littered with puzzling codes, apparent only to a noted Harvard symbologist, Robert Langdon, and to a sexy French cryptologist named Sophie—who also happens to be the dead man's granddaughter. The codes were left by the curator just before his death as a sort of treasure map to lead Sophie and Langdon to the hidden location of the mythical Holy Grail.

The unfortunate curator is not the only victim, merely the latest in a series of brutal murders—all members of a secret society called the Priory of Scion that is dedicated to guarding the secret of the grail. Sophie and Langdon must decipher a series of ciphers and riddles to keep the grail from being lost forever. Lurking in the background is the elusive figure of Leonardo himself, a supposed grand master of the Priory of Scion in his day, and the quintessential "Renaissance man": excelling not only in art but also in science, mathematics, cryptography, and engineering.

Almost as much has been written debunking Brown's book as praising it. Langdon, as Brown's mouthpiece, serves up some questionable assertions, largely based on conjecture and scant circumstantial evidence. Leaving aside the brouhaha among religious groups over the question of Jesus' divinity and supposed marriage, there are fundamental historical errors. A quick glance through any decent history book would reveal that the Merovingian kings did not found Paris. Langdon also overestimates the number of women burned as witches during the European witch hunts of the seventeenth century, giving the number as 5 million. Historical records estimate that between 30,000 and 50,000 victims were executed, but not all were women, and not all were burned.

Nonetheless, the book is a gripping page-turner, and academic standards of establishing fact usually don't apply to pulp fiction. It's normal for authors to take liberties in the name of entertainment. But Brown claims in a brief foreword that chunks of his plot are based on historical fact, and that claim makes him vulnerable to criticism. Brown did a fair amount of research, but he should have been a bit more discriminating in his sources. The only bona fide scholar cited is the gnostic gospel expert Elaine Pagels. The rest are a motley assortment of new age thinkers, goddess worshipers, and conspiracy buffs whose academic credentials are highly suspect.

Did Brown get anything right? Well, yes. There really is a so-called "divine proportion," "golden ratio," or, in mathematical terminology, simply phi. Langdon identifies its value as 1.618, although in reality it's an irrational number: a number with infinite decimal places, like pi. Phi describes an intriguing geometrical anomaly. If a line is divided into two unequal lengths, such that the ratio of the longer segment to the shorter segment is the same as the ratio of the whole line to the longer segment, then the resulting number will be something close to 1.618. Phi was first mentioned in Euclid's *Elements* around 300 BC. Geometric shapes are said to be in "divine proportion" if the ratios of their various sides closely resemble phi. The most common such shapes are the golden rectangle, the golden triangle, and the algorithmic spiral, which can be seen in any chambered nautilus shell.

The first coded message Langdon and Sophie discover is a seemingly random string of numbers: 13-3-2-21-1-1-8-5. But Sophie soon realizes the string isn't random at all, but the first eight numbers of something called the "Fibonacci sequence," though written in jumbled order. This, too, is established fact. Fibonacci—also known as

Leonardo Pisano, or Leonardo of Pisa—was a thirteenth-century mathematician who wrote the *Book of Calculation*. In it, he described a sequence in which, after the first two terms (1 and 1), each number is equal to the sum of the two previous numbers: $1 + 1 = 2$; $1+2 = 3$; $2 + 3 = 5$; $3 + 5 = 8$; and so forth. This sequence is closely related to phi. If you take each number in the sequence and divide it into the one that follows, you will get an answer that comes closer and closer to the value of phi—bearing in mind that since phi is an irrational number, the progression will go on infinitely, without ever exactly equaling phi. So 5 divided by 3 is 1.666; 13 divided by 8 is 1.625; 21 divided by 13 is 1.615; and so on.

Langdon is correct in asserting that the same sequence of numbers occurs frequently in nature; so does the golden ratio. For example, the total number of petals in most flowers is a Fibonacci number. An iris has three petals, a buttercup five, and an aster 21 petals. In a nautilus shell, the ratio of the first spiral to the next is roughly 1.618. The golden ratio can also be found in certain crystal structures, and in the swooping, spiral flight path of a falcon attacking its prey.

Salvador Dalí experimented with the concept of phi in his art; the French architect Le Corbusier advocated its use in designing buildings; and some twelfth-century Sanskrit poems have a meter based on phi, called the *matra-vrttas*. In any poem using the *matra-vrttas*, each subsequent meter is the sum of the two preceding meters, according to Mario Livio, an astrophysicist who wrote *The Golden Ratio*. One ancient Indian scholar named Gopala even calculated the series of meters: 1,2,3,5,8,13,21 ... This is the Fibonacci sequence, although Gopala didn't know that. He completed his analysis almost 100 years before Fibonacci published his treatise in 1202.

Measuring the distance from the tip of your head to the floor, and dividing that by the distance from your navel, will give you an approximation of 1.618. The classic sculpture *Venus de Milo* (c. 200 BC) is so proportioned. Bulent Atalay, a physicist at Mary Washington College in Virginia and author *of Math and the Mona Lisa*, conducted an informal survey of these proportions among 21 of his students and found that for most, the results were very close to the value of phi. All this makes phi "one *h* of a lot cooler than pi," to quote one of Langdon's students.

But there are other cited instances of phi that aren't as well substantiated; in fact, they're highly controversial. Some people, like Langdon, see the golden ratio in every aspect of life: art, music, poetry, architecture, even the stock market. For instance, the east and west façades of the Parthenon are said to form golden rectangles, that is, their ratios of length to width equal phi. Naysayers—including several mathematicians—counter that this ratio is not exactly 1:1.618, but 1:1.71. That might be close enough for a symbologist like Langdon, but for scientists the discrepancy is significant.

Ultimately the question is one of artistic intent, which is notoriously difficult to prove. There is only scant circumstantial evidence that the Egyptians consciously used phi to construct their pyramids. It's equally unlikely that Virgil deliberately based the meter of the *Aeneid* on phi. Ditto for Gregorian chants and the works of the classical composers Mozart and Béla Bártok. So is it just a coincidence that all of the above show evidence of the golden ratio? Not necessarily, says Atalay, who, as both an artist and a scientist, falls somewhere in between the two camps. He believes human beings have a natural affinity for the golden ratio, which might explain why it keeps popping up in human creations—even when it isn't intended. "If we look hard enough, we are bound to find an unlimited number of examples among man's creations in which the golden ratio occurs," he says. "One has to assess each piece separately, and guard against reading between the lines."

Things get murkier when it comes to Langdon's assertions about Leonardo da Vinci. Here too there are errors. Langdon refers to him constantly as "Da Vinci"; historians always refer to him as "Leonardo." The *Mona Lisa* is believed by most art historians to portray the wife of Francesco di Bartolomeo del Giocondo, and not to be, as Brown asserts, a self-portrait. Nor did Leonardo receive "hundreds of lucrative Vatican commissions"—he received just one, which he never completed. The Priory of Scion—purported to date back to the Merovingian kings who ruled the south of France in the sixth to eighth centuries—may never have existed, although the subject is a perennial favorite among religious conspiracy theorists. Rumors of such an order first surfaced in the 1960s with the discovery of documents in the French National Library, along with a list of supposed grand masters that includes not only Leonardo but also Isaac Newton. Most historians agree that the list is a hoax. Leonardo was a man with many secrets, but running a heretical gnostic sect devoted to the Holy Grail and goddess worship is unlikely to have been among them.

Leonardo was born in 1452, the illegitimate son of a Florentine notary, Ser Piero d'Antonio; and a local peasant girl named Caterina. Caterina was foisted off on a cowherd in a neighboring village, while Ser Piero married into a wealthy family. But he didn't abandon his son. The young Leonardo grew up in his father's household and received a solid education. When his artistic talent emerged, he was apprenticed to Andrea del Verrochio, a prominent Florentine artist. In Andrea's workshop, Leonardo learned the basics of painting and sculpture, grinding and mixing pigments, and perspective geometry. By the time he was accepted into the painters' guild in 1472, he was already making sketches of pumps, weapons, and other ingenious devices of his own design, in addition to his art.

We know that Leonardo relied on geometric perspective in both art and science. He not only read extensively on the subject but also collaborated with the mathematician Fra Luca Pacioli and the artist Piero della Francesca on a book about the golden ratio, *De divina proportione*, published in 1509. Although Leonardo never directly mentions using phi in his art, according to Atalay, "The evidence is in the grand melange of the mathematical musings, studies in perspective, formal drawings, quick sketches, and finished paintings." For example, in one unfinished painting, *Saint Jerome in the Wilderness* (now housed in the Vatican Museum), the figure of Jerome is framed within a golden rectangle. The *Virgin of the Rocks* in the Louvre has a height-to-width ratio of 1:1.62, very close to phi. *Vitruvian Man* is a study of proportion in the human male form, and contains several sets of golden rectangles. What about the *Mona Lisa*, which figures so prominently in Brown's novel? The proportions of the panel are about 1:1.45—not especially close to phi—but Atalay points out that the panel was unaccountably trimmed at some point in the painting's 500-year history, making that measurement meaningless. Within the painting, he asserts, the relationship between the woman's right shoulder and cheek and her left shoulder and cheek forms a golden triangle whose shorter sides are in divine proportion to its base. (A similar construct

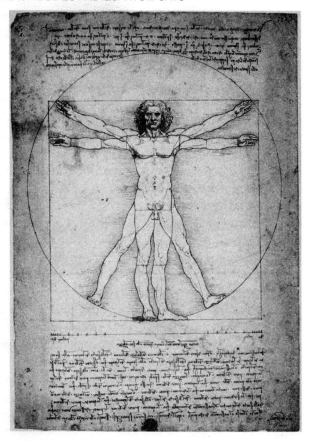

**Leonardo da Vinci's *Vitruvian Man*
contains several sets of
golden rectangles**

also turns up in self-portraits by Rembrandt.) The painting also contains a number of golden rectangles that line up with all the critical focal points of the woman's image: her chin, eye, and nose and the upturned corner of her mouth. "In the work of any other artist, we would assume these manifestations to be coincidental," Atalay writes. "For Leonardo, who seamlessly integrated mathematics, science, and art, and spent his life seeking unifying principles, perhaps not."

It's true that Leonardo saw no distinction between science and art: he argued that painting was a science, and maintained a constant cross-fertilization between the two. For instance, along with many artists and doctors of his day, Leonardo dissected cadavers to study anatomy—no doubt as quickly as possible, given the lack of refrigeration and formaldehyde in fifteenth-century Italy. This certainly informed his art, but he was equally enthralled by the structure of limbs and their dependence on nerves and joints, so his artistic skills were put to use illustrating his scientific pursuits. He made meticulous studies of horse entrails and the spinal cord of a frog, and a series of drawings that analyzed the internal dimensions of the skull.

Atalay concurs with the many historians who consider Leonardo to be the first modern scientist. In all his investigations, artistic or otherwise, Leonardo applied a technique that strongly resembles the scientific method: practical experimentation, direct observation, recording of data, and synthesis of the findings into a theoretical explanation. He called for mathematical demonstrations of scientific principles, 200 years before Isaac Newton merged math and physics forever. The eleventh-century Arab scholar Alhazen is also often credited with using an early version of the scientific method: he tested his hypothetical conclusions about the laws of reflection and refraction of light, for instance. But he and Leonardo were rarities in their respective ages.

Alas, Leonardo never published any of his findings during his lifetime, and although he produced more than 13,000 pages of notebooks (later gathered into codices), less than a third of them have survived. If he had published them, Atalay contends, "Our present level of sophistication in science and technology might have been reached one or two centuries earlier." The notebooks contain all manner of inventions that foreshadow future technologies: flying machines, bicycles, cranes, missiles, machine guns, an "unsinkable" double-hulled ship, dredges for clearing harbors and canals, and floating footwear akin to snowshoes to enable a man to walk on water. Leonardo foresaw the possibility of constructing a telescope in his *Codex Atlanticus* (1490), where he writes of "making glasses to see the moon enlarged"—a century before the instrument was actually invented. And in 2003, Alessandro Vezzosi, director of Italy's Museo Ideale, came across some recipes for mysterious mixtures while flipping through

Leonardo's notes. Vezzosi experimented with the recipes, and the result was a mixture that would harden into a material eerily akin to Bakelite, a synthetic plastic that was widely used in the early 1900s. So Leonardo may well have invented the first man-made plastic, becoming the first materials scientist.

Most of these inventions were never built. Leonardo was primarily a visionary. On his deathbed, he is said to have asked, "Has anything been done?" Something would be done, although not, perhaps, what Leonardo had in mind. Within 50 years of his death, a man named Nicolaus Copernicus would publish a book that launched an unprecedented scientific revolution, routing a system of beliefs that had dominated for more than 1000 years. The world would never be the same again.

2

Talkin' 'bout a Revolution

1542

Publication of *De revolutionibus*

Only the earth doth stand forever still,
Her rocks remove not nor her mountains meet;
Although some wits enrich with learning's skill
Say heav'n stands firm and that the earth doth fleet
And swiftly turneth underneath their feet.

—Sir John Davies, "Orchestra" (1596)

"This is a bad land for gods," says Shadow, the unlikely hero of Neil Gaiman's fantasy novel *American Gods*. Fresh out of prison, and reeling from the sudden death of his wife in a car accident a few days before, Shadow takes a job with a mysterious stranger called Wednesday. There's more to Wednesday than meets the eye: he turns out to be an old, exiled Norse god, once known as Odin the All-Father. Immigrants coming to America brought their old gods with them, joining the deities who were already there. But Americans now worship money, technology, the Internet, and celebrity. The old gods are slowly fading away from lack of belief. Most just want to enjoy their quiet retirement, though for some this means eking out a living through con games and prostitution. But Wednesday doesn't plan to go gently into that impending twilight. He is rounding up other forgotten deities and forming an army for a final showdown between the old gods and the new gods: a postmodern Armageddon.

Revolutions aren't always sudden or epic in scale, particularly if they are revolutions of ideas. In the mid-sixteenth century, the Polish astronomer Nicolaus Copernicus published a massive tome called *De revolutionibus*. Over the course of 100 years, his manuscript changed people's entire view of the universe and their place in it. Yet it seemed an unlikely catalyst: the bulk of the book was a rather dry mathematical treatise. "It was surely a text to be studied, but scarcely to be read straight through," the Copernican scholar Owen Gingerich writes in *The Book Nobody Read*. Nonetheless, copies of the first edition of *De revolutionibus* are now auctioned at more than half a million dollars.

The fuss arose over Book I, which placed the sun at the center of the known universe. This was a radical departure from the cosmological model that had prevailed for 14 centuries. Ptolemy (in Latin, Claudius Ptolemaeus), an Egyptian living in Alexandria around AD 150, was the first to gather the ideas of previous Greek astronomers into a working model for the universe. His treatise *Almagest* set forth a mathematical theory for the motions of the sun, moon, and planets. The universe was a closed, spherical space, beyond which there was nothing. The Earth was planted firmly in the center, surrounded by a set of nested spheres, or "epicycles," each of which formed the orbit for a planet, the sun, the moon, or the stars.

By the time of the Renaissance, the Ptolemaic worldview pervaded western European culture, seeping into art, literature, and music, as well as navigation, medicine, and astrology, which was then considered a legitimate science. This worldview was an aesthetically pleasing construct, and meshed nicely with the prevailing Christian theology of that era. Everything on Earth—below the moon—was deemed "tainted" by original sin, while the celestial epicycles above the moon were pure and holy, filled with a divine "music of the spheres." It became the fashion for courtier poets, like England's John Donne, to praise their mistresses as being "more than moon," and dismiss "dull sublunary lovers' love" as inferior and base. And it provided a handy rationale for maintaining the social hierarchy at all costs. "The heavens themselves, the planets, and this centre/Observe degree priority and place," Ulysses pontificates in Shakespeare's *Troilus and Cressida*. Upset the order of this so-called "great chain of being," and the result would be unfettered "chaos."

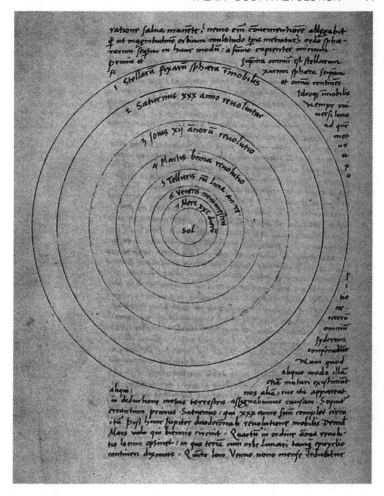

**The heliocentric
Copernican model**

From a practical standpoint, Ptolemy's approach worked well enough for making reliable predictions of planetary positions. But the predictions were good for only a few years; corrections had to be made regularly to keep the tables used for navigation, calendars, and casting horoscopes up to date. As the centuries passed, discrepancies between the calendar and key events like the spring equinox and the full moon became larger and larger. Church officials found this particularly alarming, since they based the dates of major religious holidays such as Easter on this celestial calendar. Astronomers were ready to consider an alternative model.

The man who toppled Ptolemy had an inauspicious beginning. Copernicus was the youngest of four children. When he was only 10, his father died. He was then raised by an uncle, a canon at Frauenburg cathedral. In 1492, the same year that Christopher Columbus set sail for the new world, the 19-year-old Copernicus entered the University of Kraków. He studied Latin, mathematics, geography, and philosophy, but soon his main love became astronomy. He traveled to Italy in 1496, ostensibly to pursue degrees in canon law and medicine, but after witnessing his first lunar eclipse in March 1497, he found himself drawn back to astronomy. Copernicus eventually became a canon at Frauenburg cathedral himself, and was thus assured of a comfortable income with very little in the way of actual duties. The life suited him, since he had plenty of leisure time to pursue independent study. He built an observatory in his rooms in the turret of the town's walled fortification, and diligently studied the heavens each night.

In 1514, an anonymous handwritten booklet began making the rounds among a few select astronomers, all personal friends of Copernicus, its author. Known as the "Little Commentary" (or, in Latin, *Commentariolus*), the booklet laid out his new model of the universe. The sun was at the center, with the Earth and other planets orbiting around it. This would account for the annual cycle of "movements" of the sun. He correctly determined the order of all six planets known at the time—Mercury, Venus, Earth, Mars, Saturn, and Jupiter—and concluded that the apparently changing positions of the stars are actually caused by the rotation of the Earth itself. Finally, he asserted that the apparent retrograde motion of the planets is caused because one is observing them from a moving Earth. These axioms would form the basis for *De revolutionibus*.

Historians believe Copernicus began writing his magnum opus around 1515, although it wasn't published until near the end of his life. Copernicus was his own worst critic, and he never felt the work was ready. The book might never have been published were it not for a young professor of mathematics and astronomy from the University of Wittenberg named Georg Joachim Rhäticus, or Rheticus. Rhäticus visited Copernicus at Frauenburg in May 1539. He originally planned to stay a few weeks, but he became so fascinated by the older astronomer's work that he wound up staying for two years. At his urging, Copernicus finally completed the manuscript. In 1542, Rhäticus delivered it to a printer in Nuremberg, then one of the centers of printing. But he wasn't able to stay and supervise the process, so he delegated that responsibility to Andreas Osiander, a local Lutheran theologian.

Rhäticus would bitterly regret that decision. Alarmed at the potentially heretical content of the treatise, Osiander replaced Copernicus's original preface with his own (unsigned) letter to the reader. In it, he said that the book's conclusions should be viewed not as the "truth," but as a new model for a simpler means of calculating the positions of the heavenly bodies. Rhäticus never forgave Osiander for what he viewed as a betrayal of his master's great work. According to Gingerich, who spent years tracking down rare first editions of *De revolutionibus*, the incensed Rhäticus struck out the replacement preface with a red crayon on several copies. By the time Copernicus received a copy of his life's work in 1543, he was said to be on his deathbed, having suffered a stroke that left him partially paralyzed. He died of a cerebral hemorrhage shortly thereafter.

Most of the early criticism of the book came from fellow astronomers like Tycho Brahe, a Danish nobleman born in 1546. Brahe's life is marked by two bizarre incidents. He wore a silver false nose strapped to his head, having lost part of his own nose when it was cut off in a duel. He also suffered a very strange death. Legend has it that while dining with the Danish emperor, Brahe badly needed to relieve himself midway through the meal. But it would have been rude to leave the table before royalty. His bladder burst, and he died a few days later—perhaps the only person in history to die from good manners.

Brahe's fame as an astronomer was ensured when, at age 26, he discovered a new star. He built what was then the best astronomical observatory in Western Europe, in a castle on a small island in the sea, close to Copenhagen. And he began systematically mapping the sky.

Brahe was not a Copernican; he once observed that Copernicus offended both physics and the Holy Scripture, in that order. But he admired his predecessor's "intellectual acumen," and credited Copernicus with pointing out certain gaping holes in the Ptolemaic system. Brahe tried to reconcile the two models. His own cosmology retained the Earth's position at the center of the universe with the sun circling the Earth, and the other planets circling the sun.

The criticisms had some foundation. *De revolutionibus* didn't offer much in the way of new observational data. The telescope wouldn't be developed for another 50 years, and there was only so much of the heavens an astronomer could see with the naked eye. So to compile his long catalog of the stars, Copernicus had relied on much of the same data as Ptolemy. And his primary argument for placing the sun at the center of the solar system seems to have been aesthetic rather than empirical: he reasoned that since the sun was the "lamp of the universe," it should be at the center.

There were other problems that slowed broad acceptance among astronomers. Most notably, Copernicus retained the popular sixteenth-century idea that all celestial objects moved in uniform circular motions, so his model still used circles within circles for planetary orbits. Again, it was aesthetically pleasing—why wouldn't the heavens move in perfect circles, where no corruption or decay could be found? But it was simply wrong. As the seventeenth century dawned, Johannes Kepler used Brahe's carefully collected observational data to map out the motions of the planets in their orbits. Like Copernicus, Kepler assumed that those motions would be circular, but he couldn't get the orbit of Mars to fit into any kind of circle at all. "Oh ridiculous me!" he wrote on the day he finally realized what the problem was: the orbits of the planets weren't perfect circles after all. They were elliptical, and this became Kepler's first law of planetary motion.

What Osiander and others found potentially heretical was the very notion of a heliocentric universe. The Ptolemaic worldview was seamlessly intertwined with traditional interpretations of key passages in the Bible, which church doctrine had declared to be infallible. The notion of a fixed sun orbited by the Earth and other planets seemed to contradict those passages. For instance, Psalm 104 says, "The Lord God laid the foundation of the earth, that it not be moved forever."

For many, this clearly implied a fixed Earth. The Protestant reformer Martin Luther denounced Copernicus in 1539—before *De revolutionibus* had even been published, although the "Little Commentary" was in circulation—spewing, "This fool wants to turn the entire science of astronomy upside down! But as the Bible tells us, Joshua told the Sun, not the Earth, to stop in its path." It apparently never occurred to church leaders that their interpretation of scripture might be wrong.

Christians in seventeenth-century western Europe also believed that man was made in God's image and was a superior being because he had a soul. Accepting Copernicus meant abandoning Ptolemy, and removing man from his place at the top of the cosmological chain of being. "The world had scarcely become known as round and complete in itself when it was asked to waive the tremendous privilege of being the center of the universe," Johann Wolfgang von Goethe later wrote of the implications of a heliocentric universe to seventeenth-century believers. "Never, perhaps, was a greater demand made on mankind."

Naturally, the old gods fought back. Espousing the theories of Copernicus could prove hazardous to your health in the early seventeenth century. Consider the infamous case of Giordano Bruno, an Italian astronomer who adopted the Copernican cosmology early on. In fact, he went one step farther. Bruno dared to suggest that space was boundless instead of fixed, and that the sun and its planets were just one of multitudes of similar solar systems. He even speculated that there might be other inhabited worlds, with rational beings equal or superior in intelligence to humans. Today, Bruno might be a bestselling science-fiction author, but his ideas were much too creative for the conservative Catholic Church to stomach. He was tortured, condemned, and eventually burned at the stake in 1600 for refusing to recant his beliefs.

Bruno is often held up as the first Copernican martyr, but there is some question as to whether he was executed specifically for his Copernican beliefs. Bruno was a mystic who adhered to several doctrines considered heretical at the time. He was also a bit of a hothead, becoming embroiled in disputes wherever he went. In short, he was a prime candidate for martyrdom in the charged religio-political climate of that era. The truth is that *De revolutionibus* didn't immediately raise a hue and cry among church leaders, who didn't even list it on their index of forbidden books until 1616. The church just didn't deem Copernicus's ideas worthy of notice until the Italian astronomer Galileo Galilei began making observations with a new instrument called the telescope that verified Copernican theory, lending it greater credibility and making it much more of a threat to the religious status quo.

Galileo's work—most notably the "Dialogue Concerning the Two Chief World Systems," published in 1632—brought matters to a head. Not only did his observations contradict the teachings of the church; he also audaciously reinterpreted biblical passages to conform to a sun-centered cosmology, at a time when only church elders had the authority to interpret scripture. He argued that the Bible should be used to teach people how to get to heaven, not as an instructional manual for how the heavens move. The church finally took action in 1633, dragging Galileo before the Inquisition. He was forced to his knees in front of his inquisitors and, under threat of torture and death, officially "renounced" his belief in the Copernican worldview. Unlike the unfortunate Bruno, Galileo escaped his ordeal with a lifetime sentence of house arrest.

In the end, the new gods prevailed. Science trumped the theologians, as more and more empirical evidence accumulated in support of the Copernican cosmology. By the late seventeenth century, so many new discoveries had been made with the telescope that there were hardly any astronomers who weren't Copernicans. It might be tempting to dismiss those early Catholics and Protestants for clinging to their silly assumptions about the universe when solid scientific evidence indicated otherwise. But the modern world is far from immune to faith-based superstitious beliefs. The old gods occasionally rattle their sabers, just to let us know they're still here. Their favorite postmodern adversary is the evolutionary theories of Charles Darwin.

There are almost as many camps of creationism as there are lilies of the field, but the most radical subset holds that God created the world in 4004 BC. As for the trifling fossil record indicating that the Earth is actually about 4 billion years old—well, God created that record, too, as a test of faith for believers. These so-called "young Earth" creationists reject not just evolution, but any scientific theory that contradicts their narrow interpretation of the book of Genesis, including the big bang and plate tectonics. Their mind-set is nearly identical to that of the Catholic Church toward Copernican theory: when scientific evidence contradicts their reading of the Bible, they assume that the evidence is false—not that their interpretation of scripture might be wrong.

Most creationists aren't this radical, but they have become increasingly proactive, and often litigious, particularly as regards setting science curricula in schools. The Kansas state board of education made national headlines in August 1999 when it voted 6 to 4 to eliminate the teaching of the big bang and evolution from the required scientific curricula. Waging lawsuits to gain equal classroom time for creationism hadn't worked, so creationists had quietly infiltrated the school board, which determines acceptable texts for instruction. Their victory was small, and very short-lived: a new board restored the old scientific curricula in February 2001. But the issue refuses to die. Similar battles have been waged in school districts in Ohio, Texas, Arkansas, and Alabama. Some textbooks in Alabama still contain a disclaimer stating that evolution is just a "theory."

The seventeenth-century Christians came to realize that they could embrace the Copernican system without banishing their notion of God as the creator of the heavens. And there are plenty of modern scientists who have no problem reconciling their personal religious beliefs with the precepts of science. But short of staging an ideological Armageddon, it seems that the best we can achieve is an uneasy, precarious truce between the old gods of religion and the new gods of science.

3

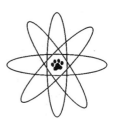

Lens
Crafters

Circa 1590
Invention of the microscope

Some 5000 years ago, ancient Assyrian scribes recorded on clay tablets the existence of magic magnifying stones that could be used to make objects in nature seem larger. These stones were actually broken shards of meteorites whose centers had fused into glass during the intense heat of their entry into the Earth's atmosphere, melting in such a way that they formed a very primitive sort of lens. Even if the Assyrians didn't know it, they were practising the earliest known form of optical microscopy.

The strange magnifying effects that so intrigued the ancient Assyrians owe less to magic, of course, than to the unique nature of light and its interaction with matter. Light is a peculiarly ephemeral substance that suffers from a chronic identity crisis. For centuries scientists have debated whether light is a particle or a wave; today we know it is neither, although it has properties of both, like an actor who moonlights as a waiter on the side, or who really just wants to direct. It travels in a straight line like a stream of particles, but it can also diffuse outward like a wave.

Anyone who has taken a basic science class knows that light is made of packets of energy, called photons. But this energy doesn't just spring fully formed from the void, like Athena from the head of Zeus; it has to come from somewhere. There are several ways of producing photons, all involving the same basic mechanism: energizing the electrons orbiting each atom's nucleus, which is usually done by applying heat in some form. A candle's flame, the sun's rays, a laser, or an incandescent lamp can each be used to raise an atom's temperature. An electron has a natural orbit (called an orbital), much like that of a satellite orbiting the Earth, but if an atom heats up (i.e., is energized), its electrons move to higher orbitals. As they cool and fall back to their normal orbits, the excess energy is emitted as photons, which our eyes perceive as light. Not all light is visible, of course; photons possess frequencies that match the distance their parent electrons have fallen. Only those frequencies in the visible portion of the electromagnetic spectrum can be seen.

All these photons travel blindly through space as electromagnetic waves. Occasionally they bump into objects in their path, and the reflections produced as they bounce off are what enable us to see the world around us. But different materials, such as glass, can absorb, reflect, scatter, or even bend light.

A lens is simply a curved piece of glass that bends incoming light—an optical effect known as refraction. Every material has its own "index of refraction," which determines the rate at which light will travel through it. That's why light travels faster through air than it does through glass. So when light enters a piece of glass at an angle—which is what happens when the glass surface is curved, as in a lens—one

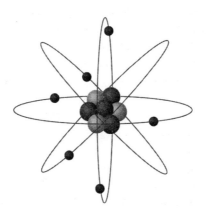

Diagram of an atom

part of the light wave hits the glass and starts to slow down before another. This causes it to bend in a given direction. As it exits the glass, the light will bend again, because parts of it enter the air and speed up before other parts. How the lens curves determines the direction in which the light bends. If the lens curves inward, light will spread out, or diffuse, like a wave; this is called a concave lens. If it curves outward, the light will bend toward the center of the lens; this is called a convex lens.

Let's say an ancient Assyrian lights a torch and stands in the center of the town square on a dark night. The blazing torch will emit light uniformly in all directions, and that light bounces off any objects in the vicinity—perhaps a statue of Ashur, the capitol city's patron god. A neighbor just happens to have a couple of makeshift shards of meteorite on hand that serve as lenses: one curving inward and the other curving outward. When he holds up the inward-curving (concave) shard, it collects the light scattered off the statue, while the outward-curving (convex) shard focuses the light, perhaps even forming a reflected image of the statue. These properties, known since antiquity, are what make lenses so useful for scientific exploration and discovery. All optical imaging techniques are based on a combination of these two types of lenses.

The oldest known man-made lens—a polished rock crystal about 1½ inches in diameter—was unearthed in the ruins of ancient Nineveh, and the Roman dramatist Seneca (4 BC?–AD 65) was said to have magnified the texts he read by peering at them through a glass globe of water. The first true magnifying glass appeared around AD 1000. This "reading stone" was a glass sphere that could be laid on top of a text to magnify the lettering. Rudimentary lenses for spectacles were being made as early as 1260; the English philosopher Roger Bacon mentions them in his *Opus majus*, published in 1268. Export records for the city of Florence list no fewer than 50 registered spectacles makers in 1400, and nearly 200 years later, eyeglasses were almost commonplace among the general populace.

Lens grinding hasn't changed fundamentally since its inception: it is still a time-consuming, methodical process, best suited to those afflicted with obsessive-compulsive disorder. In the sixteenth century, a small, round, flat piece of glass would be smeared with a "slurry"—

fine particles of sand or powder dispersed in water to keep dust to a minimum—and then rubbed to grind away the edges of the glass to the desired shape. The finer the particles, the more precise the grinding. A lens crafter would initially use a slurry with bigger particles to take a large amount of material off (modern lens makers use diamond cutting tools for this initial phase). This would grind the glass into the correct shape—convex or concave—but the particles would leave scratches on the surface, making the glass look frosted. So the lens maker would gradually work his way down to increasingly finer slurry particles to polish the surface until the glass was clear.

Each stage was followed by a meticulous washing to remove all the coarser particles before the lens maker proceeded to the next refining phase. Even just a few remaining coarse specks could leave scratches in the finished lens. The inevitable imperfections sometimes produced images in which the colors were not lined up properly, resulting in a rainbow halo around the image. That effect is known as chromatic aberration, and it occurs because different colors of light bend differently: when light moves through an imperfectly ground lens, its beam is split into separate wavelengths and produces the colors of the spectrum. And if the shape of the lens is not smoothly curved, the result will be spherical aberration: light is not refocused properly, causing blurriness.

It was a teenage eyeglass maker in Middelburg, Netherlands, named Zacharias Janssen who first came up with the concept of a microscope, although most historians believe that his father, Hans, must have had a hand in creating the instrument, given Zacharias's tender age at the time. (Some credit a fellow Dutch eyeglass maker, Hans Lippershey, with concurrent, though independent, invention of the microscope.) No record exists as to how Zacharias made his discovery, but his profession required him to experiment with lenses. At some point he must have stumbled upon the notion of combining a concave and convex lens to improve magnification and image quality, and devised a practical instrument based on his discovery. Historians are able to date the invention of the microscope to the early 1590s, thanks to the correspondence of a Dutch diplomat, William Boreel, a longtime family friend of the Janssens. Boreel wrote a letter to the French king in the 1650s—decades after the instrument's invention—detailing its origins.

A microscope uses a concave lens to gather light, and the convex lens focuses it to point, so that the image isn't blurred. A tube holds the convex lens at the proper distance from the objective lens and blocks out stray light (except for the light source needed to illuminate the specimen). No early prototypes of Janssen microscopes have survived, but a museum in Middelburg has a microscope dating from 1595, bearing the Janssen name. There are three tubes, two of which are draw tubes that can slide into the third, which acts as an outer casing. The microscope is handheld, can be focused by sliding the draw tubes in or out while observing the sample, and when extended to the maximum is capable of magnifying images up to 10 times their original size.

The term "lens" was coined in sixteenth-century Italy and supposedly derives from the Latin *lens*, which means "lentil," since early lenses resembled that legume. Alas, by modern standards, these rough prototypes were also about as effective as a lentil in terms of their ability to collect light and clearly magnify objects. Making single lenses for magnifying glasses or spectacles was difficult enough. Microscopes required two complementary lenses (one convex, the other concave), doubling the inevitable defects that crept in during the grinding process. Illumination sources were also limited to candles and oil lamps, so the images were dark. Because of these shortcomings, Janssen's invention wasn't an immediate success. More than half a century would pass before the instrument was sufficiently improved to find widespread use among scientists. The scientist Henry Power of Yorkshire was the first to publish observations made with a microscope, and in 1661 Marcello Malpighi used a microscope to observe the capillary vessels in the lungs of a frog, effectively proving the English physician William Harvey's theory of blood circulation .

An English draftsman turned scientist named Robert Hooke was among the first to make significant improvements to the basic design. Although he relied on an instrument maker in London, Christopher Cock, to actually build his microscopes, Hooke enjoyed a reputation as one of London's finest makers of precision scientific instruments. He had a passion for astronomical instruments like the telescope, and for clocks. In fact, as a child he once examined the various parts of a brass clock, then used what he learned to build his own working model out of wood. Hooke was skilled at lens grinding—he claimed that the long hours he had spent bent over a lathe since age 16 caused his pronounced stoop later in life—and this helped him to achieve much-needed improvements in magnification. But his instruments were still deficient by modern standards: they were plagued by blurred, dark images. Hooke tried passing light generated from an oil lamp through a glass filled with water to better illuminate his specimens. It was an ingenious approach, but it did not result in better images.

Still, Hooke could see details undetectable to the naked eye, and his observations formed the basis for his magnum opus, *Micrographia*, published in January 1665. It caused an immediate sensation, on a par with Stephen Hawking's best-selling *A Brief History of Time* in terms of its impact on the average person on the street. For scientists, it provided a wealth of new data, illustrated by 58 stunning engravings. Thanks to his early training as a draftsman, Hooke was able to render in painstaking detail the dark, blurred images produced by his instrument. He was also the first to apply the term "cell" (from the Latin *cella*, for a small apartment, or monk's cell) to describe the features of plant tissue, after observing cork from the bark of an oak tree under his microscope.

A large part of the book's popular appeal lay in its accessible writing style and ample illustrations, which offered an astonishing new perspective on common everyday objects—everything from carrots and a piece of silk fabric, to insects and bacterial molds. The enlarged depiction of a flea, rendered in nightmarishly vivid detail, no doubt kept many a child awake after bedtime. No less a luminary than the diarist Samuel Pepys claimed to have pored over the volume until the wee hours of the morning, declaring it "the most ingenious booke that I ever read in my life." Pepys was so inspired that he went out and bought his own microscope and later joined the Royal Society, eventually becoming its president in 1684. *Micrographia* literally changed the way people looked at the world. Hooke was even immortalized on the stage in a 1676 production called *The Virtuoso*, by playwright Thomas Shadwell. The farce drew much of its pratfall humor from the character of Sir Nicholas Gimcrack, an amateur scientist who "spent two thousand pounds on microscopes to find out the nature of eels in vinegar, mites in cheese, and the blue of plums, which he has subtly found out to be living creatures."

It was a Dutch draper, Antoni van Leeuwenhoek, who made further improvements to the instrument. He built more than 500 microscopes—only a handful have survived—and he is sometimes popularly credited with inventing the microscope, even though his work came some 75 years after the Janssens first experimented with the microscopic properties of lenses. Leeuwenhoek, a great admirer of Hooke's *Micrographia*, began making his own observations around 1673. His instruments were the best of his era in terms of magnification: he achieved magnifying power up to 270 times larger than the actual size of the sample, using a single lens that gave clearer and brighter images than were achieved by any of his colleagues. His microscopes were little more than powerful, handheld magnifying glasses; it was his skill at lens grinding and the care he took in adjusting the lighting that increased the magnifications. The single lens was mounted in a tiny hole in the brass plate making up the body of the instrument, and the specimen was mounted on a sharp point just in

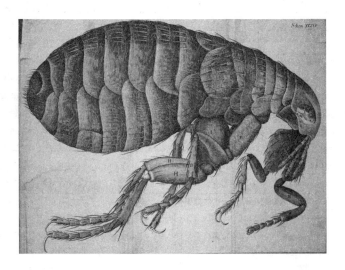

**Hooke's drawing of a flea,
from *Micrographia***

front of it. The position and focus could be adjusted by turning two screws. The entire instrument was only 3 to 4 inches long.

Leeuwenhoek used his microscopes to study protozoans found in pond water, animal and plant tissues, mineral crystals, and fossils. He discovered such microscopic creatures as nematodes, as well as blood cells, and was the first to see living sperm cells of animals. By 1683, he had turned the instrument on himself to study the plaque between his teeth, and also observed teeming hordes of bacteria in the mouths of two elderly men who had never cleaned their teeth in their lives. Apart from what this tells us about the sad state of oral hygiene in the seventeenth century, it was the first observation of living bacteria ever recorded. He even experimented with using the ovum of a cod and the corneas of dragonflies as biologically derived lenses, succeeding in generating clear images of a candle flame with the latter. He was still dictating new observations on his deathbed in 1723, at more than 90 years of age.

Today, scientists continue to invent a plethora of ingenious new microscopic techniques, but the basic principles remain the same. It is now possible to observe living cells, as well as individual atoms, at unprecedented resolutions—as small as the thickness of a human hair. And in the spirit of the early pioneers of microscopic research, scientists at Florida State University have whimsically turned their advanced instruments on common everyday objects like that all-American staple, burgers and fries (and all the trimmings), detailing thin sections of wheat kernel, onion tissue, starch granules in potato tissue, and crystallized cheese proteins.

Scientists have even figured out how to replace light with sound waves to build acoustic microscopes, and have speculated about the possibility of constructing flat lenses with no special need for shaping, using so-called "left-handed" materials. Such materials have what is called a negative index of refraction, enabling them to deflect light rays in the direction opposite what optics would normally predict. All this should convince even an ancient Assyrian that modern-day science works its own kind of magic.

4

Good Heavens

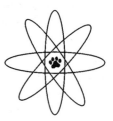

In 1990, after eight years in the making, the Hubble space telescope was launched amid much fanfare and with great expectations. At long last, astronomers could bypass the dust, air currents, and water vapor in the Earth's atmosphere that had hindered clear imaging of distant stars with ground-based telescopes. Now they could take pictures from deep space itself and, they hoped, learn more about the earliest cosmic structures that began forming moments after the big bang.

Alas, poor Hubble got off to a rocky start. Once it was in orbit, astronomers found they could not focus the instrument, which was producing fuzzy, indistinct images instead of images with the expected unprecedented clarity and fine detail. Scientists scrambled to find the problem. It turned out that the primary mirror lens had been ground to the wrong dimension. Hubble's mirrors are so smooth, and so precisely shaped, that if its primary mirror were scaled up to the diameter of the Earth, the biggest "bump" on the surface would only be 6 inches high. So even though the defect was less than 1/50 the size of a human hair, the resulting distortion was equivalent to what a 15-year-old with 20/20 vision would see if he tried on his grandfather's bifocals.

In short, Hubble needed contacts. Red-faced, NASA scientists devised a replacement "corrective lens" to fix the defect. It consisted of several small mirrors that would intercept the beam of light from the flawed mirror and correct the defect, then relay the corrected beam to the scientific instruments, in much the same way that eyeglasses or contact lenses correct vision by intercepting and correcting the angle of incoming light by which we perceive objects around us. Once the "contact lens" was implemented, Hubble began producing breathtaking images of distant celestial objects, and NASA breathed a collective sigh of relief. But the incident illustrates the critical importance of having the right lens prescription. Despite all its advanced technology and dizzying array of sophisticated instruments, Hubble fell victim to the same difficulty that has plagued eyeglasses, microscopes, and telescopes since their inception: defects in the lenses.

In Philip Pullman's fantasy novel *The Amber Spyglass*, an exiled physicist, Mary Malone, ingeniously constructs her own primitive telescope lenses out of a lacquer made from tree sap, fitting them at either end of a bamboo tube. The underlying scientific principles haven't changed significantly since the instrument was invented in the 1590s. Historians generally agree that the telescope originated in the Netherlands. The first real telescope was invented by a German-born eyeglass manufacturer named Hans Lippershey, who opened a spectacles shop in Middelburg—the same town where the inventors of the microscope, Hans and Zacharias Janssen, lived. According to contemporary accounts, Lippershey noticed two children playing with lenses in his shop, and overheard them exclaim that when they looked through two lenses, a weather vane on a nearby church appeared to be larger and clearer. After trying the trick himself, Lippershey realized the possibilities and placed a tube between the lenses to make what he called a *kijker* ("looker").

A telescope operates on optical principles similar to those involved in the microscope, but with some key differences, since the purpose of one is to examine the very small while the other is intended to examine faraway objects. The reason we have difficulty seeing objects that are far away is that they don't take up sufficient space on the eye's retina for the retinal sensor to detect them. If the eye were larger, it could collect more light from the object to create a brighter image, and then magnify part of it so that it stretches over more of the retina. Essentially, that is all a telescope is: an extension of the human eye. Unlike a microscope, which needs to gather only a little light from a small area of an illuminated sample, a telescope must gather large amounts of light from a dim, distant object. Thus it needs a large concave (objective) lens to gather as much light as possible, which a smaller convex lens then brings to a focus.

Lippershey may have built the first known telescope, but it didn't take long for other inventors to catch up. Two simultaneous patent applications for similar instruments appeared in October 1608 in the Netherlands; a third inventor apparently developed a telescope around the same time and attempted to sell it at the Frankfurt fair. These early designs were all refractor telescopes, so named because they used a combination of concave and convex lenses to refract, or "bend," light. They were able to collect light and magnify object images by two or three times the original size. News of the invention spread like wildfire throughout Europe.

By April 1609, citizens could purchase "spyglasses" in spectacles makers' shops in Paris. These spyglasses were largely a novelty item; no doubt people with voyeuristic leanings used them to spy on their neighbors, much as bored New Yorkers use binoculars today. Within four months, spyglasses were also available in Italy, and a prominent Italian scientist in Pisa named Galileo Galilei first heard of this marvelous new instrument for "seeing faraway things as though nearby" in a letter from his colleague Paolo Sarpi, who had witnessed a demonstration in Venice. Galileo was the first to recognize its potential use in astronomy. Unsatisfied with the performance of the available instruments, he duplicated and improved the invention, even learning how to grind his own lenses. By late October, he was ready to turn his telescope toward the heavens.

The first object Galileo studied was the Earth's moon. He did this toward the end of 1609, when Jupiter was closest to Earth, and hence the brightest object in the night sky, apart from the moon itself. He noted on January 7, 1610, that Jupiter appeared to have three fixed stars nearby. Intrigued, he looked again the following night, expecting Jupiter (which was then in retrograde) to have moved from east to west, leaving the three stars behind. Instead, Jupiter seemed to have moved eastward. Puzzled, Galileo returned to the formation repeatedly, observing several significant details. First, the stars appeared to be carried along with the planet. Second, they changed their respective positions. Finally, he discovered a fourth such star.

By January 15, Galileo concluded that the objects were not fixed stars, but small moons that revolved around Jupiter. This was an era when astronomers believed that everything in the universe revolved around the Earth. The notion of a planet having its own moons was in direct opposition to that belief. If Jupiter had four orbiting moons, then the Earth could not be the fixed center of the universe. This was the first empirical support for Copernicus's theory that the sun, not the Earth, was at the center of the solar system. Galileo published this startling observation in his book *Sidereus nuncius (Starry Messenger)*, which appeared in Venice in March 1610, guaranteeing his fame and ensuring his place in scientific history.

These and many other observations made with early telescopes eventually helped topple the misguided (and somewhat narcissistic) notion that everything revolved around the Earth. Sadly, Galileo did not live to benefit from the coming ideological revolution; he was convicted of heresy by the Inquisition in 1633. He spent the remainder of his life under house arrest, making what observations he could with the instruments at hand, and writing letters to colleagues and loved ones. He died on January 8, 1642. Not until October 31, 1992—350 years after his death—was the church's condemnation of Galileo officially revoked by Pope John Paul II.

Refracting telescopes like the ones built by Galileo were sufficient to see details in planets and nearby stars, but astronomers quickly realized that there were limits to the amount of light a refracting telescope could collect, and hence how far into space it could see. Around 1680, Isaac Newton developed the first reflecting telescope, which used mirrors instead of lenses to reflect rather than bend light, producing much brighter images. For that reason, reflecting telescopes soon replaced refracting telescopes among astronomers, although both types are still available today. In a reflecting telescope, a primary mirror in back of the tube collects light and reflects it to a focus via a small flat secondary mirror. The light is then deflected through the side of the tube to the eyepiece, forming an image. The brighter light made the reflecting telescope much more useful for observing faint, deep-sky objects like galaxies and nebulae. The nature of the latter was a subject of much speculation among astronomers, because nebulae appeared to be little more than clouds of gas and dust in interstellar space.

It would be another 100 years before the amorphous blobs of the mysterious nebulae were resolved into clusters of stars, thanks to the vastly improved reflecting telescopes built by the German-born astronomer William Herschel. When he was in his early twenties, Herschel emigrated to England, where he made a comfortable living as a musician and chorister. At age 35, he became an avid astronomer. He initially rented a small reflecting telescope to observe the heavens, but found it insufficient to make the kind of detailed observations he desired. Since he lacked the funds to purchase a larger instrument, he decided to build his own. Good-quality optics were expensive, so he decided to make his own mirrors, too, and purchased equipment for pouring the metal, tools for grinding and polishing, and some ground disks.

His initiative paid off in ways he never expected. The large reflecting telescopes that he constructed far surpassed in size those of his contemporaries, providing an unprecedented view of distant objects in the night sky. His largest telescope, a 40-foot reflector, remained the world's largest such instrument for the next 50 years. Herschel discovered two of Saturn's moons with this telescope in 1789, but the mirror needed frequent repolishing, the instrument weighed 2 tons, and the focusing tube was difficult to handle. Thus most of his recorded observations were made with a smaller reflecting telescope. His skill with optics eventually earned him a considerable supplementary income, since telescopes continued to be a popular novelty item for amateur stargazing, even if only the nobility could afford Herschel's pieces. He sold instruments to the king of Spain, the Russian court, and the Austrian emperor.

On the night of March 13, 1781, Herschel observed an unusual disk-shape object, which he initially thought was a comet. For the next few months he kept an eye on the object's movements, and he discovered that the object's orbit was nearly circular and lay well beyond the orbit of the planet Saturn. He concluded that it was a planet, which he called *Georgium sidus* ("George's star") in honor of the English king. Fortunately, other astronomers intervened and changed the name to the far more mellifluous Uranus, after the mythological god of the skies. Herschel's discovery—the first new planet found since antiquity—brought him immediate celebrity and earned him a knighthood plus a royal pension.

To make his observations, Herschel used a technique he called "star gauging": making sample counts of the stars in the field of view of his telescope to map out how they were arranged in space. He would train his telescope on a particular point and watch what crossed the field of view in a thin strip of the visible sky. He did this while standing on a ladder, calling out descriptions of the objects he observed to his sister Caroline, who stood below and dutifully recorded the information. On each successive night, he changed the telescope's position to observe another thin strip of sky, then another, and so forth. Over a period of about 20 years, with Caroline's help, Herschel was able to map out all of the night sky—at least as much of it as was visible in Great Britain. During that time he methodically cataloged some 2500 nebulae in addition to more than 90,000 stars. In 1845, an aristocratic amateur astronomer, Lord Rosse, erected a 72-inch telescope dubbed "Leviathan" and used it to discover the spiral nebulae.

Many modern telescopes incorporate elements of both the refractor and the reflector designs and are called compound telescopes. Hubble is a very large compound telescope. It has a long tube that is open at one end and mirrors to gather and focus light to its "eyes"—an

Herschel's 40-foot telescope

elaborate system of cameras and sensors that enable it to detect differ-ent types of light, such as ultraviolet and infrared. Light enters the telescope and bounces off a primary mirror to a secondary mirror, which redirects the light through a hole to a focal point just beyond the primary mirror. Smaller mirrors distribute the light to an imaging spectrograph, a device that acts like a prism. It separates the collected light into the component colors and analyzes the different wave-lengths. The cameras use charge-coupled devices (CCDs) instead of photographic film to capture light—the same array of photosensitive diodes used in digital cameras. The captured light is stored in onboard computers and relayed to Earth as digital signals, and these data are then transformed into images.

Astronomers can glean a lot of useful scientific information from the Hubble images. The colors of light emitted by a celestial object form a chemical fingerprint of that object, indicating which elements are present, while the intensity of each color tells us how much of an element is present. The spectrum can also tell astronomers how fast a celestial object is moving away or toward us through an effect called the Doppler shift. Light emitted by a moving object is perceived to increase in frequency (a blue shift) if it is moving toward the observer; but if the object is moving away from us, it will be shifted toward the red end of the spectrum. (A popular bumper sticker plays on this concept by declaring, "If this sticker is blue, you're driving too fast!") And while interstellar gas and dust can block visible light from celestial objects, even with a space-based telescope, it is still possible to detect infrared light, or heat, from objects hidden in clouds of dust and gas.

Because it takes so long for light to travel from distant objects to the Earth, when we look deep into space we are also looking back in time. Hubble made headlines again in March 2004 when it produced the deepest look yet into the visible universe, showing galaxies that formed between 400 and 800 million years after the big bang—a time that is very brief on the grand cosmic scale, because the big bang is believed to have occurred 12 billion to 14 billion years ago. The relative smattering of photons detected—compared with the millions detected per minute from closer galaxies—were so ancient that they had begun their journey across the universe before the Earth even existed, limping exhaustedly into Hubble's field of view billions of years later.

When NASA launches the new James Webb space telescope in 2011, we may be able to get even closer to the birth of the universe and obtain a more complete understanding of how it evolved. That is because the new instrument's larger mirror and improved optics will be able to capture more light much more quickly, imaging in a few hours objects that now may take Hubble weeks to record. The James Webb space telescope also has a large shield to block light from the sun, the Earth, and our moon, which might otherwise cause the instrument to heat up and interfere with the observations. So the new instrument will continue to push the cosmological envelope, peering through the interstellar dust to record light left over from the birth of stars and other planetary systems.

5

Gravity's Guru

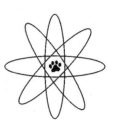

July 1687
Publication of Newton's *Principia*

> *And this is the sole mortal who could grapple,*
> *Since Adam, with a fall, or with an apple.*
>
> —Lord Byron, *Don Juan*, Canto X

Sibling rivalry can be a brutal thing. One scene in the humorously macabre hit film *Addams Family Values* depicts Wednesday and Pugsly on the roof of the family's decrepit Gothic mansion, their new baby brother and a cannonball in hand. The perpetually poker-faced Wednesday poses a scientific hypothesis to her brother: If Baby Pubert weighs 10 pounds, and the cannonball weighs 20 pounds, which will hit the stone walkway below first if they are dropped simultaneously from the roof?

Pugsly ponders this conundrum and guesses, "The cannon-ball?"

"Very good," his sister says, deadpan. "But which one will bounce? There's only one way to find out." And she and Pugsly unceremoniously drop both baby and cannonball from the roof.

Wednesday's fiendish experiment is a clever nod to Galileo Galilei's legendary experiment from atop the leaning tower of Pisa, in Italy. But the Addams children got the answer wrong. Or rather, they guessed right, but for the wrong reason. Their experiment is interrupted when the family patriarch, Gomez, leans out a window and catches the baby in mid-fall. Had the baby continued on course, the cannonball would indeed have hit the ground before Pubert. But its earlier arrival would have had nothing to do with its relative weight.

Wednesday and Pugsly were hardly the first to assume that heavier objects will fall faster than lighter ones. The notion dates back to Aristotle, and anyone who has dropped a rock and a piece of paper at the same time has observed that the rock will hit the ground first. But this is due not to how much either object weighs, but to friction caused by the presence of air. The air exerts more friction on Pubert than on the cannonball because the baby has more surface area. This slows Pubert's fall, and the cannonball hits the ground first. Had the Addams siblings conducted their experiment in a vacuum, there would have been no air resistance and both baby and cannonball would have hit the ground at the same time, despite the 10-pound difference in weight.

This is precisely what Galileo observed in the early seventeenth century: all objects fall at the same rate, no matter what their mass, and the force that makes them fall is gravity. Galileo didn't really drop cannonballs from the leaning tower of Pisa, although he often used the notion as an illustrative example; the story is apocryphal. But he did experiment extensively with gravity. Rather than drop balls of different weights from a height, he rolled them down a gentle incline so that they would move much more slowly, making their acceleration easier to measure. (The balls were similar in size, but some were made of iron and others of wood.) Lacking an accurate clock, Galileo reportedly timed the balls' travel with his pulse. And he quickly realized that no matter what the incline, the balls would travel at the same rate of acceleration.

Wednesday also confuses mass with weight. It's an important distinction, since weight fluctuates with gravity whereas mass is constant. Mass simply denotes how much matter an object contains. It is a measurement of the total number of subatomic particles (electrons, protons, and neutrons) an object contains. When people on the South Beach diet claim they've lost weight, they really mean they've lost mass by shedding some of those pesky excess subatomic particles that make up body fat. Weight is the amount of force exerted on a given object by gravity. A man may weigh 170 pounds on Earth, but his weight decreases as he travels from the Earth to the moon, because the Earth's gravitational pull lessens the farther he moves away from it. His mass will remain the same, because the number of atoms that make up his body is the same.

Although many scientists throughout history experimented with and pondered the mysteries of gravity, it was Isaac Newton who built upon the insights of his predecessors to define its characteristics mathematically. The son of a yeoman farmer in Lincolnshire, England, Newton was born in 1642 (the same year Galileo died), premature and so small that hardly anyone expected him to survive. His mother, Hannah, once said she could fit the baby Isaac into a quart pot. Like many scientists, he showed an early affinity for mechanics and mathematical problems. As a child he was enthralled by the construction of a new mill in the nearby town of Grantham. He studied the principles that governed the various gears, levers, and pulley wheels, and he built his own water mills and windmills. He even built a 4-foot water clock out of a wooden box in his attic room.

As a young man, Newton attended Cambridge University, earning his undergraduate degree in science and math in 1665. His graduate studies were interrupted by an outbreak of plague in Cambridge. Students and professors alike fled the city, and Newton returned home for a year, until the panic (and danger) had passed. But he was far from idle. In that one year, he laid the groundwork for revolutionary ideas that would change the course of scientific history.

Like many astronomers of his day, Newton was intrigued by the mysterious force that kept the moon and planets firmly fixed in their respective orbits. Legend has it that one day during his exile to escape the plague, he was ruminating on the orbital problem in his orchard, when he saw an apple fall from a tree. (The tree blew over in 1820 during a storm, but bits of it were grafted onto trees on an old estate in Belton, ensuring that at least some part of it would survive.) Newton theorized that there had to be a force acting on the apple to cause its downward acceleration. And if that force reached the top of the highest tree, it probably also extended sufficiently out into space to affect the orbit of the moon. The only force that could account for this was gravity, a term which was by then in fairly common use among scientists. It derives from the Latin *gravis*, meaning "weighty" or "heavy." He concluded that as the moon orbits, it falls toward the

Earth. But it is also moving forward, so it does not hit the Earth but it falls past, and gravity curves it around into a regular orbit. The moon is continuously "free-falling" in its path around the Earth.

Newton compared the phenomenon to firing a cannon from atop a mountain so high that its peak is above the Earth's atmosphere. The cannonball will eventually fall to the ground, because gravity directs it toward the center of the Earth. The more charge is used with each firing of the cannon, the faster the ejected cannonball will move, and the farther it will travel before falling to Earth. If enough charge were used, the cannonball would travel so far that it would circle around the curve of the Earth, always falling within the gravitational field but never actually hitting the ground. The cannonball would be in orbit around the Earth.

In *Addams Family Values*, Wednesday and Pugsly are sent against their will to Camp Chippewa, where they run afoul of the relentlessly cheerful counselors and nauseatingly wholesome offspring of the "privileged elite." When they and their fellow outcasts stage a revolt during a camp pageant, they catapult one of the spoiled rich girls into the lake. If she were catapulted with sufficient force from a platform above the Earth's atmosphere, she could theoretically be sent into orbit around the Earth, just like Newton's hypothetical cannonball. As shallow, vain, and diet-obsessed as her nitwit clique, the girl could take comfort in the fact that objects in orbit are weightless—not because they are beyond Earth's gravitational field but because they are in a continuous free fall.

Over the next two decades, Newton expanded and codified his insights into three basic physical laws. First, a body at rest will remain at rest, and a body in motion will remain in motion unless an outside force—such as friction or a collision with another solid object—intervenes. This is the law of inertia. We know that there are two kinds of energy: potential and kinetic. Potential energy, such as food in the human body or gasoline in a car's engine, is waiting to be converted into power. Kinetic energy is energy of motion. When Pugsly practices archery at Camp Chippewa, he increases potential energy by making the bowstring taut; that potential energy is then converted into kinetic energy when the arrow is released. According to Newton's first law, Pugsly's arrow will travel indefinitely until it is overpowered by an opposing force—friction, gravity, or, in this instance, the last remaining American bald eagle.

Kinetic energy increases with the velocity squared, which is a physicist's way of saying that if one car, for example, is traveling twice as fast as another, it will have four times the energy. Similarly, the higher you lift an object, the more potential energy the object gains, and the greater the resulting kinetic energy. The cannonball Wednesday drops from the roof will speed up by 9.8 meters per second for each second that it falls. So if it falls for 5 seconds, it will reach a speed of 49 meters per second—a rate of acceleration equivalent to a car going from 0 to 60 miles per hour in less than 3 seconds. In another attempt to eliminate their new sibling rival, Wednesday and Pugsly drop an anvil from the third floor down the stairwell to the foyer, where baby Pubert is contentedly gurgling on the floor. Because it is dropped from a significant height (and the baby is rescued just before the anvil hits), the anvil crashes through the floor to the basement below. Had the children been on the ground floor and simply dropped it from where they stood, the anvil, having accumulated less kinetic energy, might only have cracked the floor.

Acceleration is caused by the application of force. According to Newton's second law, the more force is applied to an object, the greater the rate of acceleration. When baseball player Roger Clemens throws a pitch, he applies a force to the baseball that causes it to speed up. Gomez Addams applies a similar accelerating force when he throws knives at Uncle Fester or hurls darts at a board held by Lurch. This effect must be balanced against the mass of an object: the more mass an object has, the greater the force required to get it moving and the more slowly it will accelerate. The Addams family's hearse will accelerate more slowly with the entire family onboard than it would if it held just Lurch. These two factors make it possible to calculate how much force is required to move any given object. Newton boiled the calculation down to a single equation: force equals mass times acceleration ($F = ma$).

According to his third law, for every action there is an equal and opposite reaction. For example, firing a gun accelerates the bullet forward, and this force is countered by the recoil action, which causes the gun to snap backward. This happens because of conservation of momentum. The total amount of momentum in the universe is constant and cannot change. The Addamses' hearse will stay in place because gravity exerts a downward force upon it, while the ground exerts an equal and opposite upward force on the tires. The hearse begins to move forward only when the accelerator is pressed down and the engine applies enough extra energy to overcome gravity. Once the hearse is in motion, the air exerts a counterforce (aerodynamic drag) in the opposite direction. The amount of drag increases as the vehicle gains speed, requiring more energy to overcome the resistance and leaving less energy available for acceleration. The hearse reaches "cruise speed"—the point where it can accelerate no more—when the driving force is equal to the aerodynamic drag force.

Newton conceived much of this while he was still in his twenties, and he set up numerous intricate experiments involving weights, pulleys, and pendulums to test his mathematical predictions, all duly recorded in his notebooks. For instance, he suspended various substances of different weights—gold, silver, lead, glass, sand, salt, wood, water, and wheat—in two identical wooden boxes hung from 11-foot cords, and timed how long it took for the different substances to fall. That was how he arrived at his notion of a universal gravitational constant. But it wasn't until 1684, when he was visited by the astronomer Edmond Halley (after whom Halley's comet is named) that Newton was inspired to refine his calculations and compile them in a book. This book summed up everything he had discovered about gravity.

Always a workaholic, Newton toiled like a man possessed to complete his manuscript, taking meals in his rooms, rarely venturing out, and often writing while standing up at his desk. His notorious absentmindedness grew even worse: it was not uncommon for him to leave a dinner party to get more wine, only to be found, hours later, slaving over an unfinished proof, both wine and friends forgotten in his feverish enthusiasm. Over the course of three years, he wrote *Mathematical Principles of Natural Philosophy* (known by its shortened Latin name, the *Principia*), a massive three-volume work that is one of the most influential scientific books ever written. The famous eighteenth-century mathematician Joseph-Louis Lagrange described the *Principia* as "the greatest production of a human mind."

Knighted in 1705, Newton was careful to give due credit to the work of those who came before him, insisting that if he had seen farther than others, it was only because he stood on the shoulders of giants. "I do not know what I may appear to the world," he once wrote in his journals. "But to myself, I seem to have been only like a boy, playing on the seashore and diverting myself, in now and then finding a smoother pebble, or a prettier shell than ordinary, whilst the great ocean of truth lay all undiscovered before me." The man who had stood on the shoulders of giants died at Kensington in London on March 20, 1727, and is buried at Westminster Abbey.

Newton's work on gravity and acceleration would remain unchallenged for more than 200 years, until a former patent clerk turned physicist named Albert Einstein, in his theory of general relativity, made an equally radical intuitive leap. Newton correctly calculated the strength of gravity, but scientists didn't really know how it worked. According to Newton, gravity is a force that acts immediately on an object. Celestial bodies move in straight lines except when gravity pulls them into curved orbits. His model correctly predicts how objects fall and planets move. But that approach doesn't take into consideration that there will always be a time delay equivalent to at least the speed of light, which is the cosmic speed limit. When we "see" the sun, what we are actually seeing is light that left the sun roughly 500 seconds ago; we only seem to be seeing the sun "now." Nothing can travel faster than light, so gravity can't act on objects instantaneously.

Einstein resolved the problem by dispensing altogether with the need for a gravitational force that acts through space: instead, he saw gravity as being embedded in the cosmos. Gravity, he insisted, isn't a force at all; it's a curvature in the fabric of space-time, and objects must follow that curvature. The presence of mass or energy doesn't affect objects on Earth directly; it warps the space around it first, and the objects move in the curved space. According to general relativity, the Earth always travels in a straight line.

The presence of the sun curves space, and thus the Earth appears to be moving in an elliptical orbit. Imagine that Wednesday and Pugsly are holding a sheet, stretched out taut horizontally. If Baby Pubert is placed in the center, his mass will cause a depression in the sheet. If Thing then places an apple on the edge of the sheet, it will roll down the slope toward the center. The depression can't be seen by someone looking straight down at the sheet from above, so it appears that the apple was pulled by an invisible force. In Einstein's cosmos, the sun's mass causes a similar depression in the fabric of space-time, and the planets in their orbits act like the apple.

This was an interesting theory, but how could Einstein prove it? Babies and cannonballs (or apples) would not do the trick. Instead, Einstein turned to light, which also follows the curvature of space. Einstein predicted that when a ray of light passes near a massive body, like the sun, the ray will bend because the sun's mass warps the surrounding space. This deflection could be measured when the sun's own light was blocked during an eclipse. He even calculated how much the light should be deflected. During a solar eclipse in May 1919, two separate scientific expeditions—one on an island off west Africa, and the other in Brazil—photographed stars near the eclipsed sun. They found that the starlight was deflected, just as Einstein had predicted.

Scientists still use Newton's equations to predict the effects of gravity—the cannonball will always hit the ground before the beleaguered Pubert, assuming the presence of friction—but it was general relativity that explained the "what" and "how" of the equations, and changed the way we think about objects in space and time.

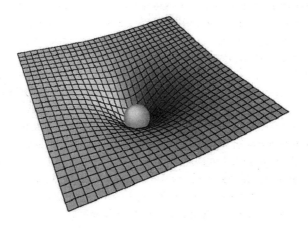

Gravity according to Einstein

6

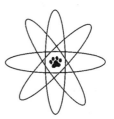

Drawing Down the Fire

June 1752
Franklin's kite experiment

In a 1995 episode of *The X-Files*, the FBI agents Fox Mulder and Dana Scully investigate the strange case of a teenage boy who is struck by lightning, miraculously survives, and then becomes a human conductor, able to channel the millions of volts in a lightning strike without coming to harm, and to manipulate electric fields. It is the only talent this otherwise unremarkable high school dropout possesses, and it defies scientific explanation, as befits a TV show devoted to the paranormal. But one refrain repeated throughout the episode is true: scientists still know very little about how lightning actually works.

They do know quite a bit about electricity. The word comes from the Greek *elektor*, "beaming sun"; and the Latin *electrum*, meaning amber, a fossilized tree sap that has hardened to the consistency of stone over millions of years. Around 600 BC, the Greeks noticed a peculiar property of amber: when rubbed against a piece of fur, for example, it would attract particles of dust, feathers, and straw. But the strange effect remained little more than a curiosity until about 1600, when an English physician named William Gilbert began studying the reactions of magnets and amber and realized that other objects could also be made "electric." He concluded that friction created this "electricity."

The secret of how friction gives rise to electricity lies in atomic structure. Negatively charged electrons spin around the nucleus of the atom. The nucleus is made up of positively charged protons and of neutrons, which, like the Swiss, remain neutral. In stable atoms, such as carbon, the positive and negative charges are balanced; that is, there is the same number of protons and electrons. Deep down, all atoms want to be balanced; it's their natural state. But some atoms have electrons that are more loosely attached, and these so-called "free" electrons are able to move, or "flow," between atoms of matter, creating a current of electricity. The promiscuous little electron charge is passed from atom to atom, the way children try to pass on "cooties" by tagging schoolmates (or siblings) they dislike.

These unstable, or charged, atoms are called ions. An atom that gains electrons becomes negatively charged, whereas an atom that loses electrons becomes positively charged. Free electrons that have been kicked out of the atomic "house" simply move around until they find an atom with a positive charge to give them a home. At the atomic level, at least, opposites really do attract: positively charged atoms are always on the prowl for negatively charged free electrons to balance themselves out. It's yin and yang, a perpetual atomic singles bar. And if they have enough energy, they'll even knock off electrons already orbiting an atom to make room for themselves.

There's an easy way to visualize conductance. If there is a "live" downed power line on the road a few feet from where you happen to be standing, chances are you'll survive if you keep your distance. The atoms that make up the dry asphalt are stable and non-conductive, and the electron charges in the atoms of the power line have nowhere to go. But what if the road is wet, with a thin sheet of water or ice running along the surface? Unless you happen to be wearing thick rubber soles, you'll probably be electrocuted by the high-voltage currents coursing through the power line. Water is highly conductive, and if that downed line touches the water's surface, it transfers electrons to the water. The sudden influx of high-energy electrons will knock off the electrons orbiting the water's atoms. These must find a new home, so they in turn knock off electrons from neighboring atoms, and so forth. The cascade continues until the electrons hit a nonconducting substance, like rubber or glass, in which the electrons are too firmly attached to be easily displaced. Metal is also an excellent conductor. A teenage boy in Ang Lee's film *The Ice Storm* (1997) is electrocuted while leaning against a guardrail along the side of the road. When a downed power line touches the guardrail, the current whizzes along the metal strip and ultimately into the boy's body, killing him instantly.

That's the science behind current electricity. Drag your feet on the carpet and then touch a conducting surface, like a metal doorknob. The resulting shock is static electricity. The shuffling causes our bodies to pick up extra negatively charged electrons. Touching something with a positive charge, like metal, causes the electrons to "jump" across the small gap from our fingers to the object. Lightning is a particularly spectacular form of static electricity: its high voltage and high intensity make a lightning strike lethal to human beings, rather than merely annoying. Lightning occurs because clouds become negatively charged as the water droplets inside rub up against each other during the natural process of evaporation and condensation, when moisture accumulates in the clouds. This charge seeks out something with a positive charge—the ground, ideally—and the lightning is the "spark" closing the gap between the two.

Americans of all ages are familiar with the story of Benjamin Franklin and his famous experiment to determine whether lightning was in fact an electrical current: he attached a metal key to a kite during a thunderstorm to see if the lightning would pass through the metal. Contrary to popular belief, Franklin, a self-educated amateur scientist, wasn't the first to conduct this experiment successfully, although he certainly designed it.

Franklin was born in Boston on January 17, 1706, the tenth son of a soap maker named Josiah Franklin, who would eventually father 17 children with his three successive wives. He could afford to send young Benjamin to school for only one year.

Since the boy loved to read, he was apprenticed to his brother James, a printer, at age 12. He helped compose pamphlets, set type, and even sold the printed products on the street. Three years later, James started the *New England Courant*, Boston's first real newspaper, which carried original articles and editorials rather than simply reprinting news from abroad. However, relations between the two brothers were not good, and James even beat Benjamin occasionally. In 1723 Franklin ran away, eventually landing in Philadelphia. He initially found work as an apprentice printer, then borrowed money to set up his own printing business, which soon became successful. Franklin also purchased his first newspaper, the *Pennsylvania Gazette*, and in 1733 founded *Poor Richard's Almanack*. By 1749 he had retired from business to focus on scientific experiments and inventions, concentrating on the mysterious phenomenon of electricity.

Franklin was the first to demonstrate that electricity consisted of a common element he called "electric fire," and that it flowed like a liquid, passing from one body to another without being destroyed. He studied charged objects, noted how sparks jumped between them, and concluded that lightning was merely a massive electric spark, similar to sparks produced from a charged Leyden jar, an early form of capacitor used by experimenters to help store electricity. A Leyden jar is nothing more than a glass jar partially filled with water, with a conducting wire projecting from its cork. (It was invented in Leyden, the Netherlands, in 1745 by the Dutch physicist Pieter van Musschenbroek, who nearly electrocuted his best friend during the first successful test run. The jar was charged by exposing the protruding end of the wire to an electric spark generated by friction.) To test his theory about the nature of lightning, Franklin published a paper proposing an experiment with an elevated rod or wire to "draw down the electric fire" from a cloud, with the experimenter standing in the protection of an enclosure similar to a sentry box.

Before Franklin could put his proposal into practice, a Frenchman named Thomas François d'Alibard read the published paper and then used a 50-foot vertical rod to draw down lightning in Paris on May 10,1752. One week later, M. Delor repeated the experiment in Paris; he was followed in July by an Englishman, John Canton. One unfortunate physicist did not fare so well. Georg Wilhelm Reichmann attempted to reproduce the experiment, but a glowing ball of charge—the first documented example of ball lightning—traveled down the string, jumped to his forehead, and killed him instantly. A few days later, however, a Russian chemist named Mikhail Lomonosov performed the same experiment successfully.

Franklin was unaware of these other experiments when he undertook his own version during a thunderstorm in June 1752, on the outskirts of Philadelphia. He built a kite out of two strips of cedar nailed together in the shape of a cross or X, with a large silk handkerchief forming the body, since silk could withstand the wet and wind of a thunderstorm. He attached a key to the long silk string dangling from the kite; the other end of the silk string was attached to a Leyden jar by a thin metal wire. He stood under a shed roof to ensure that he was holding a dry portion of the kite string (if it got wet, it would be conductive). Franklin's son, then 21, assisted him in raising the kite, and then they settled down to wait.

Just when Franklin was beginning to despair, he observed the loose filaments of twine "stand erect," indicating electrification—the same thing that happens when static electricity sets one's hair on end. He pressed his knuckle to the key and was rewarded with an electric spark. The negative charges in the key were attracted to the positive charges in his body, causing the spark to jump from the key to his hand. This proved conclusively that lightning was static electricity. It's a common misconception that Franklin was actually struck by lightning during the experiment. In fact, the spark he observed resulted because the entire experimental apparatus was surrounded by a very strong electric field. He proved his point, but—lacking the resilience of the boy in *The X-Files*—he was very lucky to have experienced only a small fraction of lightning's power. A bolt of lightning would easily have killed him.

Franklin's experiment is a small-scale version of what happens when lightning strikes the Earth. Clouds form as water evaporates into the air. As more and more water droplets collide inside a cloud, they knock off electrons. The ousted electrons gather at the lower portion of the cloud, giving it a negative charge. Eventually the charge becomes so intense that electrons on the Earth's surface are repelled by the growing negative charge and burrow deeper into the Earth. The Earth's surface becomes positively charged, and hence very attractive to the negative charge accumulating in the bottom of the cloud. All that is needed is a conductive path between cloud and Earth, in the form of ionized air—another byproduct of the collision process. There are usually many separate paths of ionized air stemming from the cloud, called "step leaders" because they work their way down toward the Earth in stages, following not a straight line but the path of least resistance. It might seem as if a straight line would be the path of least resistance; after all, it is usually the shortest distance between two points. But that is not the case.

Think of the paths of ionized air as roads in a city like Washington, D.C. Cars, like lightning, can travel only along those roads. Tourists in the nation's capital are often flummoxed by the odd layout of the city. A road will run in a neat, straight line toward the center of town and then, all of a sudden, stop. An enormous bureaucratic edifice is blocking the way. The cars must go around the obstacle via a side street to rejoin the original road, which continues on the other side. Lightning encounters similar obstacles as it travels from the clouds to the Earth. It will hit a patch of air that isn't ionized, and is thus nonconductive, and must find an alternative, conductive route. As the lightning works its way down, objects on the Earth's surface respond to the strong approaching electric field by "growing" positively charged streamers. These reach out to the cloud, like Michelangelo's Adam reaching out to God on the ceiling of the Sistine Chapel, stretching upward to bridge the gap. When the two charges finally meet, current jumps between the Earth and the cloud, producing lightning.

Impressed with the power of lightning, Franklin went on to develop the lightning rod as a protective measure. This is a pointed metal rod attached to the roof of a building. The rod is connected to a piece of copper or aluminum wire, which is in turn connected to a conductive grid buried in the ground nearby. Lightning rods don't actually attract lightning; rather, they provide a low-resistance path to the ground. Franklin understood that lightning can strike and then seek out a path of least resistance by jumping around to nearby objects that provide a better path to the ground. A lightning rod provides a safe option for the lightning strike to choose. To prove how destructive lightning can be, he built a "thunder house" with a can inside filled with flammable gases, connected by wire to a metal weather vane on the roof. He used a Leyden jar to make a spark, which traveled from the roof to the can, igniting the gases. The force of the explosion was sufficient to knock the roof off the house.

Franklin also invented a related device called "lightning bells" that would jingle when lightning was in the air. On top of his house, he erected a lightning rod, which drew lightning down. The rod was connected to a bell, and a second bell was connected to a grounded wire. Every time there was an electric storm, the bells would ring and sparks would illuminate his house, greatly exasperating his wife, Deborah.

**Franklin's
thunder house**

According to legend, she became so irritated by the constant jingling that she wrote to her husband while he was in London asking how to disconnect the experiment.

Scientists continued to delve into the nature of electricity long after Franklin had moved on to focus on other inventions and on his diplomatic career. In the latter half of the eighteenth century, the Italian anatomist Luigi Galvani constructed a crude electric cell with two different metals and the natural fluids from a dissected frog, after noticing that the leg of a dissected animal could be made to move through the application of an electric charge. This resulted in the development of the electric battery in 1800 by Alessandro Volta, who modified Galvani's cell by substituting other metals and replacing the biological material with wet pasteboard, all stacked together. It also led to much wild speculation among scientists about the potential of electricity to reanimate dead tissue. Experiments were reportedly conducted in secret on the bodies of recently hanged criminals—providing the plot for Mary Shelley's novel *Frankenstein* (1818), and later inspiring Robert Louis Stevenson's short story "The Body Snatcher" (1884). But the most important discovery about electricity was still to come. It turns out that electricity has a "soul mate" in the form of another natural force called magnetism, and the two are so closely joined that they are virtually inseparable. Together, they give rise to the entire spectrum of light and form the underpinnings for most of modern technology—truly a marriage made in heaven.

7

Trompe
l'Oeil

In 1999, the painter David Hockney broached a controversial theory that early artists—including fifteenth-century masters like Jan Van Eyck—had used optical aids even before scientists began using them. This, Hockney believes, accounts for the sudden dramatic improvements in the rendering of perspective and proportion in fifteenth-century art, as well as the remarkably realistic appearance of portraits beginning around the same time. His book *Secret Knowledge*, published in 2000, expanded on his theory. It was met by a storm of criticism from art historians but generated renewed interest in the use of optical aids by artists, past and present.

Hockney's theory isn't all that outlandish. Unusual optical effects have long fascinated both scientists and artists. Take, for example, the camera obscura (Latin for "dark room"), the precursor of the pinhole camera. In its simplest form, the camera obscura is little more than a small hole in a shade or a wall, through which light passes from a sunlit garden, for example, into a darkened room, projecting an inverted image of the scene onto a wall opposite the hole. An artist could tack a piece of sketch paper to the wall, trace the outlines of the subject, then complete the painting.

The phenomenon results from the linear nature of light and is evidence of the "particle" side of its dual personality. Light reflects off each point of an object and travels out in all directions, and the pinhole acts as a lens, letting in a narrow beam from each point in a scene. The beams travel in a straight line through the hole and then intersect, so that light from the bottom of the scene hits the top of the opposite wall, and light from the top of the scene hits the bottom, producing an upside-down image of the outside world. The visual effect can be quite striking. "When I first saw an image projected like this, I just thought I was seeing God," the artist Vera Lutter told *The New Yorker* in March 2004. Another artist, Adam Cvijanovic, delights in a naturally occurring camera obscura that appears regularly in his apartment in Brooklyn, when the light and angle are just right.

The German astronomer Johannes Kepler coined the term "camera obscura" in the early seventeenth century, but by then the phenomenon had been known for millennia; in fact, it is perhaps the oldest known optical illusion. Some form of camera obscura was most likely behind a popular illusion performed in ancient Greece and Rome, in which spectral images were cast upon the smoke of burning incense by performers using concave metal mirrors—hence the expression "smoke and mirrors." The earliest mention is made by the Chinese philosopher Mo-tzu (or Mo-Ti) in the fifth century BC; he called it a "collecting place," or "locked treasure room." There are also references in the writings of Aristotle, who noticed that during a partial eclipse, the openings between the leaves of a tree cast images of the sun onto the ground, but he failed to come up with a satisfactory explanation for the effect.

The eleventh-century Arabian scholar Alhazen of Basra is said to have had portable camera obscura tent rooms for solar observation. This was the earliest practical application for the device; it meant that one did not have to look directly at the sun and endanger one's eyesight. Alhazen, the son of a civil servant in what is now Iraq, had devoted himself to science. He devised a plan to construct a dam to control flooding of the Nile River in Egypt and pitched his idea to Egypt's ruler, Caliph al-Hakim, who granted him a commission to do just that. But al-Hakim, who would go down in history as the "mad caliph," was mentally unbalanced; he didn't provide sufficient funds, materials, or labor to complete the project, and he wasn't inclined to be sympathetic about failure. Alhazen, fearing for his life, pretended to be mad himself until the caliph died in 1021. This entailed being under self-imposed house arrest, and during that time Alhazen started exploring the phenomenon of the camera obscura.

Alhazen started with a darkened room with a hole in one wall, and hung five lanterns in the room next door. He noticed that five "lights" appeared on the wall inside the darkened room. Because the lanterns and the hole were arranged in a straight line, he correctly concluded that light travels in straight lines. And even though the light from all five lanterns traveled through the hole at the same time, it didn't get mixed up in the process: there were still five separate "lights" on the wall inside the darkened room. Since Aristotle, the common assumption had been that the eye sent out rays of light to scan objects. Alhazen determined that light was reflected into the eye from the things one observed. He also recorded the laws of reflection and refraction, correctly attributing the effects to the fact that light travels more slowly through a denser medium. His treatise on optics, *Kitab al-manadhirn*, was translated into Latin in the Middle Ages—one of only a handful of his more than 200 works that survived.

In 1490 Leonardo da Vinci described a camera obscura in his notebooks, having built a small version to test his own theories about the workings of the human eye. In 1544 the Dutch scientist Reinerus Gemma-Frisium used one to observe a solar eclipse. It was seventeenth-century Neapolitan scientist Giambattista della Porta who first suggested that the camera obscura would be a useful aid for artists, enabling them to make quick sketches of a scene with the correct perspective. Image quality was improved with the addition of a mirror placed at a 45-degree angle to reflect the image down onto a viewing surface. By the nineteenth century there were not only portable versions but entire rooms serving as entertainment. One such room still exists as a tourist attraction on the cliffs at Ocean Beach in San Francisco.

The Dutch painter Jan Vermeer (1632–1675) is believed to have used the camera obscura. The American etcher and lithographer Joseph Pennell first speculated on this possibility in 1891, citing as evidence the "photographic perspective" of certain paintings by Vermeer. Others have noted that Vermeer seems to reproduce certain "out-of-focus" effects, such as his treatment of highlights: the reflection of light from shiny surfaces. As for where Vermeer may have learned

illum in tabula per radios Solis, quam in cœlo contingit: hoc eſt, ſi in cœlo ſuperior pars deliquiū patiatur, in radiis apparebit inferior deficere, vt ratio exigit optica.

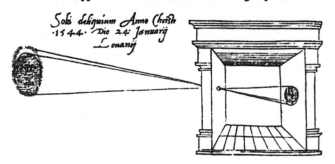

Sic nos exactè Anno .1544. Louanii eclipſim Solis obſeruauimus, inuenimuſq; deficere paulò plus q̃ dex

**Illustration of a
camera obscura**

about optics, his contemporary the microscope maven Antoni van Leeuwenhoek lived in Delft just a few streets away from Vermeer. Some scholars have even suggested that Leeuwenhoek was the model for Vermeer's two paintings with scientific themes: *The Astronomer* and *The Geographer*. In both, the subject bears a curious resemblance to known portraits of Leeuwenhoek.

More recently, Philip Steadman—a professor of architecture and town planning at University College London and the author of *Vermeer's Camera*—took a new approach to study Vermeer's paintings. He worked backward from the conventional method of artists for setting up perspective views, reconstructing the geometry of the spaces depicted. On the basis of his analysis, he concluded that as many as 10 paintings by Vermeer present the same room. Steadman maintains that his perspective reconstructions make it possible to plot the positions in space of the theoretical viewpoint of each of the 10 paintings. But he didn't just rely on mathematics. He built a one-sixth scale physical model of the room in question, outfitted it with a photographic plate camera in place of Vermeer's camera obscura, and then created photographic simulations of the paintings, testing light and shadow—all of which Vermeer faithfully reproduced in paint, lending credence to Steadman's theory. The BBC built a full-size reconstruction for a television film on the subject. It cast a full-size image of *The Music Lesson* onto a translucent screen, bright enough to show up on film, and certainly sufficient to have served as a drawing aid for Vermeer.

Later artists, such as Ingres (1780–1867), most likely used the camera lucida, an optical aid invented by William Wollaston in 1807. Wollaston was an English physician whose independent research encompassed chemistry, physics, astronomy, botany, mineralogy, crystallography, physiology, and pathology, as well as optics. Having made a comfortable fortune from his invention of a technique for making platinum in a more malleable form for industry, he abruptly abandoned his medical practice to devote more time to scientific research. The fact that he was becoming partially blind may have spurred his interest in optics.

Wollaston is described in contemporary accounts as being pleasant in appearance and very polished and refined in manner, but he lived alone and worked in rigid seclusion. His work in optics began in 1802, when he developed the refractometer, an instrument for determining refractive indexes to gain the optimum reflections. He also designed a special type of prism, made from calcite or quartz, that acted as a polarizing beam splitter; and he worked extensively in designing lenses. All this would ultimately feed into his design for the camera lucida.

Wollaston's camera lucida consisted of an extendable telescopic tube in three pieces, containing a reflecting prism and sighting lens, mounted on a stick that could be attached to a drawing table or surface. The device was more portable than the camera obscura, and the darkened room was no longer needed; the artist could work in direct sunlight. The prism had four sides, with one at a right (90-degree) angle, two at a 67.5-degree angle, and one at a 135-degree angle. These numbers were not arbitrary, but the result of careful experimentation on Wollaston's part to obtain all the critical angles of reflection. (Many of us learned the maxim in high school physics: the angle of incidence is equal to the angle of reflection.) The prism was deliberately shaped and oriented so that the rays of light from the scene were reflected

**Wollaston's camera lucida
in action**

twice within the prism before reaching the eye. This way, the eye sees the image the right way up, rather than inverted as with the camera obscura. When the stand is adjusted so that the prism half-covers the pupil of the eye, the draftsman has the illusion of seeing both the object he wishes to draw—which is reflected through the prism—and its outlines on the drawing board.

The camera lucida requires considerable skill to use, since it doesn't actually project an image of the subject onto paper; the image seems to appear on the drawing surface only when the artist looks into the prism. A slight movement of the head will cause the image to move also, disrupting the accuracy of the tracing. The modern version is remarkably similar in concept and structure, using a half-silvered (semitransparent) mirror to achieve one of the two required reflections, so that the viewer can see both the reflected scene and also the view directly through the mirror, as if it were plain glass. This also makes it easier to use, since the positioning and stillness of the eye are not as critical.

To test his hypothesis about the early use of optical aids, Hockney recruited the physicist Charles Falco, a professor of optics at the University of Arizona, who systematically analyzed measurable distortions in early paintings. "The images themselves supply the evidence, if you know how to read them," says Falco. For instance, in Lorenzo Lotto's late Italian Renaissance painting *Husband and Wife* (c. 1543), the geometric keyhole pattern of the carpet loses focus as it recedes into the painting, and oddly, there are two vanishing points clearly visible in the detail of the fabric's border. Had linear perspective been used, the pattern would have receded in a straight line, the single vanishing point corresponding to a single viewpoint. Instead, there is a kink in the pattern, which then continues in a slightly different direction. Hockney and Falco see this as evidence that Lotto used some sort of lens to project and trace the pattern of the cloth, but then found that he could not keep it all in focus at the same time; so he refocused the lens to complete the back portion of the cloth, changing the vanishing point, which he painted "out of focus" in an attempt to camouflage the process.

The earliest evidence of the use of optics that Hockney and Falco could find is in a sketch of Cardinal Niccolò Albergati made by the Flemish artist Jan van Eyck in 1431. The subject's facial features are perfectly rendered. And although the finished painting is 41 per cent larger than the sketch, when the latter is enlarged and laid over the painting, many features line up exactly: forehead, right cheek, nose, mouth, eyes, and even laughter lines. Falco insists that to have scaled up the sketch so precisely, van Eyck must have used an optical aid.

But neither the camera obscura nor the camera lucida existed as a common sketching device until the late sixteenth century, after Lotto and van Eyck had stopped painting. Hockney was at an impasse until Falco chanced to comment that a concave mirror has all the optical qualities of a lens and can also project an image onto a flat surface. Small pocket mirrors and spectacles first appeared toward the end of the thirteenth century, and Hockney reasoned that van Eyck and his contemporaries most likely owned such "mirror lenses" and might have used them in their work.

Hockney conducted his own experiment to see if the idea was plausible. He created a small makeshift window in his studio and asked a portrait subject to sit just outside in the bright sunlight. He then placed a small shaving mirror opposite the window and darkened the room to improve clarity. The result was an inverted image of his subject projected onto a sheet of sketching paper mounted next to the window, from which the outlines of the portrait could easily be sketched.

Hockney's theory still has its detractors. Among the most vocal is David Stork, a consulting professor of electrical engineering at Stanford University who maintains that there are persistent flaws in the theory. Take the carpet keyhole pattern in Lotto's painting. Falco and Hockney's analysis is based on the assumption that the pattern would have been symmetrical in real life. But Stork conferred with carpet historians, who said that the patterns in these so-called "Lotto carpets" are asymmetric, and the degree to which the pattern deviates matches the perspective anomalies in Lotto's painting.

Or consider the chandelier depicted in van Eyck's painting *Portrait of Arnolfini and His Wife* (1453). Hockney has argued that van Eyck used a "mirror lens" to trace the outlines of the chandelier to get it in proper perspective. The problem, says Stork, is that the chandelier isn't in perfect perspective. He and Antonio Criminisi, a researcher for Microsoft in Cambridge, England, analyzed a digitally scanned image of the chandelier in three dimensions, rotating the six arms of the chandelier and overlapping them to see if they aligned. They didn't. Stork even hired a British realist artist, Nicholas Williams, who painted a similar chandelier by eye that wasn't in perfect perspective but was still more accurate than van Eyck's. To Stork, this is conclusive evidence that van Eyck painted the chandelier by eye.

Falco isn't convinced by these arguments. "Everything that Stork has said is wrong, misleading, or irrelevant—or all three," he says. For instance, he says that Stork has fundamentally misunderstood how early artists would have used lenses: not to painstakingly trace an object's image and render it exactly, as in a photograph, but to capture points of perspective and a rough outline of certain objects that might otherwise be more difficult to capture on canvas. This would include van Eyck's ornate chandelier, as well as patterns on fabrics following folds, folded draperies, and architectural curves. Once the outlines were in place, the artist would turn the canvas over and fill in the details later by eye. (A cursory glance through Hockney's book confirms Falco's assertion that this point is made repeatedly.) "Paintings are not photographs," says Falco. "If you assume that the chandelier is a photograph, that's a wrong assumption, and you're going to reach the wrong conclusion."

The controversy is unlikely to be resolved any time soon: like the debate on the use of the golden ratio by artists, Hockney's theory is difficult to prove or disprove conclusively. But there is one point on which all three men can agree: the use of optics by any artist does not constitute "cheating," or in any way diminish artistic achievement. Such misconceptions "cloud the discussion," according to Stork. For Hockney—who has used a modern camera lucida himself—such aids are merely ingenious tools. They did not produce impressive results in the hands of less skilled artists. As for Falco, "I always thought van Eyck was a genius," he says. "I underestimated him."

8

Rules of Attraction

1820
First evidence of electromagnetism

Among the most powerful mutants in the *X-Men* comic books and films is the archvillain Magneto, who has the power to manipulate electromagnetic fields. He first discovers his power as a boy in a Nazi concentration camp, forcibly separated from his parents, who are sent to the gas chamber—a detail that makes the character all the more compelling. Now an embittered and vengeful old man, Magneto uses his power to stop moving trains, break into high-security vaults, and even induce unnatural mutations in "normal" humans, with horrific consequences.

Magneto and his unique ability may be fictional, but electromagnetism is a real force to be reckoned with. As its name implies, it describes the unique relationship between electricity and magnetism, which are essentially two sides of the same coin. Legend has it that around 900 BC, a Greek shepherd named Magnes walked across a field of black stones, which pulled the iron nails out of his sandals and the iron tip from his shepherd's staff. The region became known as Magnesia. This story may or may not be true, but it is known that in 1269, Petrus Peregrinus of Picardy, France, discovered that natural spherical magnets made of iron-rich ore, called lodestones, could align needles between two pole positions, forming the first rudimentary compasses—although no one was sure exactly how or why they worked.

Magnetism arises from the continual movement of charged electrons in atoms. That movement creates an electrical current, and whenever electricity moves in a current, a magnetic field is created. The reason some materials are magnetic while others are not has to do with how the electrons are ordered in a given material. In most materials, the electrons send their magnetic field in different directions, effectively canceling each other out. In magnetic materials, such as most metals, the electrons can be aligned in one direction and exert a pull together. As with electricity, the scientific study of magnetism did not begin in earnest until 1600, when William Gilbert, court physician to Queen Elizabeth I, discovered that the Earth itself is a giant magnet, like one of Peregrinus's stones. This solved the mystery of how a compass worked. In fact, much of the impetus for magnetic research in the eighteenth and nineteenth centuries came from the need for a compass that would still be accurate near the polar regions, to aid explorers searching for the fabled Northwest Passage.

Magnetism was also the basis for the "miracle cures" touted by the German physician Franz Anton Mesmer, who developed the theory of "animal magnetism" and used hypnosis (known as "Mesmerism"). Mesmer, who practiced in Vienna, enjoyed brief celebrity in Paris in the 1780s; in Mozart's opera *Così fan tutti*, a character is cured by a Mesmer magnet. Never one to ignore a fad, however fleeting, Marie Antoinette was an enthusiastic devotee. When she wasn't conceiving of ever more elaborate hairstyles, or milking cows with her ladies-in-waiting in a misguided attempt to "get back to nature" after reading Rousseau (her own eighteenth-century version of the television reality show *The Simple Life*), she could be found at one of Mesmer's famous "magnetic seances." An assistant would place the "patients" together in a magnetic tub filled with glass powder and iron filings, and lull them into a relaxed state with the music of a glass harmonica, played behind curtains covered with astrological symbols. Then Mesmer himself would appear, clad in a long purple robe—à la Albus Dumbledore in the Harry Potter books—and would touch the patients with a white wand, supposedly sending them into a trance from which they would awaken fully cured.

It was a very convincing act, but eventually Mesmer aroused the suspicion of King Louis XVI, who, perhaps irritated by his wife's slavish attachment to the physician, appointed a special committee in

1784 to look into Mesmer's methods. The committee included Benjamin Franklin (by then the reigning expert on electricity), Antoine Lavoisier (who discovered oxygen), and the infamous Dr. Guillotin, who would ultimately lose his own head to the machine that bears his name. (The French revolutionaries did not want for irony.) Several "demonstrations" of animal magnetism were carried out at Franklin's home in Paris by one of Mesmer's chief disciples, with disastrous results for Mesmer. When blindfolded, patients were unable to tell whether the "mesmerizer" was present or not, and the committee concluded that Mesmer's "cures" were the combined result of terrific showmanship and the patient's imagination. Discredited, Mesmer retired to obscurity in Switzerland.

Until 1820, scientists assumed that electricity and magnetism were two distinct forces. That was the year when a professor at Copenhagen University named Hans Christian Ørsted uncovered the first evidence that the two were, in fact, the same force, although it would be several decades before that theory was conclusively proved. Ørsted was demonstrating the basics of magnetism, and how electrical current can heat a wire. For a magnet, he used a simple compass needle mounted on a wooden stand. He noticed that every time the electric current was switched on, the compass needle moved—though so slightly that his audience didn't even notice. A few months later he repeated the experiment in his own laboratory, setting up a fine wire connected to a battery, and laid the wire over the glass cover of a large compass. Once again, when he connected the wire to the battery, the compass gave a small jerk.

Ørsted's discovery caused a sensation among scientists all over Europe, who began conducting their own research in hopes of identifying the exact nature of this unique relationship between the two forces. The man who would contribute the most to unlocking the secrets of electromagnetism was Michael Faraday. The son of an English blacksmith, Faraday had only a rudimentary education. At age 14, he was apprenticed to a bookbinder, a choice of profession that enabled him to read voraciously, which he did, particularly about the natural sciences. In 1813, toward the end of his apprenticeship, a friend gave Faraday a ticket to hear the eminent scientist Humphrey Davy lecture on electrochemistry at the Royal Institution. (Davy

**Ørsted experiments
with electromagnetism**

discovered the chemical elements barium, strontium, sodium, potassium, calcium, and magnesium.) Faraday was so taken by the presentation that he asked Davy to hire him. Davy initially declined, but shortly afterward sacked his assistant for brawling, and hired Faraday as a replacement. Some have said that Faraday was Davy's greatest discovery.

Faraday lost no time getting up to speed as a new member of the scientific community, and when news of Ørsted's discovery reached England, the young assistant enthusiastically plunged into designing his own experiments, hoping to prove beyond doubt that not only electricity and magnetism but all natural forces were somehow linked. He made three significant discoveries. In 1821, he discovered electromagnetic rotation, which converts electricity into mechanical motion by using a magnet. This is the underlying principle behind the electric motor. Ten years later, he succeeded in showing that a jiggling magnet could induce an electrical current in a wire. The principle of the dynamo or electromagnetic induction, as this setup was known, became the basis of electric generators, which convert the energy of a changing magnetic field into an electrical current. So Faraday proved that electricity gives rise to magnetism, and vice versa.

Finally, to see how light would be affected by a magnet, Faraday placed a piece of heavy leaded glass on a magnet's poles and then passed light through the glass. Normally, light vibrates in all directions at once; but polarized light passing through some sort of filter, like a lens, to reflect, bend, absorb, or scatter light waves, vibrates in a single direction. That was the purpose of the glass. When Faraday turned on the electromagnet, he found that the polarization, or direction, of the light had rotated slightly. Clearly, the light had changed when it passed through the magnetic field. This is called the magneto-optical effect, and it suggested that magnetism was related not just to electricity but also to light.

From his more than two decades of experiments, Faraday concluded that magnetism was the center of an elaborate system of invisible electric lines of force that spread like tentacles throughout space just as roots of trees branch through soil. He believed that magnetic attraction occurs because every such tentacle exerts a pull along its length. Magnetic repulsion occurs when the lines of force push against each other. Where these lines are packed closely together, such as near a magnet's poles, the forces are strong; where the tentacles are more widely spaced, the forces are much weaker.

Physics has always maintained a delicate balance between theory and experiment. Despite his reputation as a brilliant experimentalist, Faraday was largely self-educated. Moreover, he was mathematically illiterate, unable to comprehend the equations in the pioneering papers on electromagnetism, or to provide a mathematical basis for his ideas. So his theory about electric lines of magnetic force wasn't taken seriously by other scientists, with one exception. The physicist and mathematician James Clerk Maxwell, of Cambridge, thought Faraday might be on to something, and in 1862 he set about providing the missing mathematical equations to support Faraday's ideas.

Maxwell envisioned electricity as a fluid and merged this notion with Faraday's electric lines of force to come up with the concept of one continuous wave. Maxwell theorized that every time a magnet jiggled or an electric current changed, a wave of electromagnetic energy would spread out into space like a ripple on a pond. He called this wave the "flux." He calculated the speed of those hypothetical waves, and lo and behold, it was the same as the speed of light. So Faraday was correct: electricity, magnetism, and light were all connected. In fact, among modern scientists, light is synonymous with electromagnetic radiation. Ironically, the set of formulas we know today as "Maxwell's equations," which are the basis for this unification of forces, are actually the work of a telegraph operator and amateur mathematician, Oliver Heaviside, who reduced Maxwell's original 20 equations to 4.

Maxwell's theories remained hypothetical until a German student at the Berlin Academy of Sciences named Heinrich Hertz took on the formidable challenge of experimentally detecting electromagnetic waves. It took him eight long years, by which time Maxwell had died (at the relatively young age of 48) and Hertz was a professor at Karlsruhe. In 1888, Hertz hit upon the idea of putting a block of paraffin wax between the electrode plates of a capacitor, which he then rapidly charged and discharged. If electromagnetic waves were present, sparks would appear across a small gap in the detector loop. He was stunned when sparks appeared not just in the detector but over the entire experimental apparatus, providing dramatic proof for Maxwell's equations. The frequencies of electromagnetic waves are measured in hertz, in his honor.

Electromagnetism is more than just an intriguing scientific phenomenon; it can be a very powerful tool. Almost all the modern technologies we take for granted—telephones, televisions, radio, radar, X-ray machines, lasers, and electronics, to name a few—rely on electromagnetic effects. In the film *Ocean's Eleven* (2001; a remake of the original version of 1960), a team of con artists use a physics device—called simply "the pinch"—to help them rob a bank vault filled with house earnings from three casinos. (They don't have Magneto.) The pinch generates an intense electromagnetic pulse that blacks out the city's power grid for a few moments, disabling the vault's security systems in the process. As far-fetched as the idea may seem, such a device actually exists, albeit in a form different from the one portrayed in the film. (Nor do any of the scientists who work with the real-life machine resemble the film's stars, George Clooney and Julia Roberts—more's the pity.)

The fictional pinch fits inside a van. Sandia National Laboratories in New Mexico needs an entire building to house the world's largest electrical generator, called the Z pinch, which, as it fires, creates lightning-like tangles of startling color for a few billionths of a second. These flashes can be captured on film. The generator gets its name from the fact that an initial burst of electricity creates a magnetic field which compresses, or "pinches," a gas of charged particles. This generates intense X rays, and a surprisingly weak electromagnetic field—nowhere near sufficient to knock out a vault security system, in case anyone is contemplating a similar foray into crime. The real goal of Sandia's Z pinch is to achieve fusion: a potentially endless source of energy. The movie's device has more in common with so-called "e-bombs," electromagnetic pulses capable of wiping out the circuitry of all electrical devices within range, effectively crippling all technology in a modern city. Fortunately for casinos everywhere, scientists have yet to produce a sufficiently powerful e-bomb. The only known source of that strong an electromagnetic pulse is a nuclear explosion—in which case a power outage or bank robbery would be the least of anyone's worries.

Another thing electromagnetism can't do is cure illnesses. Despite the debunking of Mesmer in the eighteenth century, magnetic therapy is still widely touted as a remedy for aching joints and muscles, even though most magnetic therapy kits are little more than common refrigerator magnets, which are incapable of any measurable physical impact. This therapy produces essentially the same placebo effect that made Mesmer so successful. But electromagnetism has yielded at least one genuine medical application: magnetic resonance imaging (MRI). MRI exploits the natural magnetism of hydrogen atoms in the human body by using large, powerful magnets and pulses of radio waves to provide unparalleled detailed images of almost any body part, such as the brain, or how blood flows in the body. MRI systems might not cure any illness, but they can often diagnose what ails you.

9

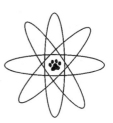

Ingenious
Engines

Circa 1840s
Babbage's analytical engine

How often do engineers get to be heroes? In the 1995 film *Apollo 13*, three astronauts find themselves stranded in space when an explosion in one of the fuel tanks irreparably damages their spacecraft. In a pivotal scene, the team leader Jim Lovell must perform critical calculations by hand, since the guidance computer has been shut down. The slightest error in calculation could mean the difference between life and death. So Lovell asks a team of NASA engineers at Mission Control—armed with the stereotypical slide rules, horn-rimmed glasses, and bad comb-overs—to check his numbers. They do so, giving a series of "thumbs up" as they complete their figures, verifying the accuracy of Lovell's on-the-fly calculations.

The Apollo 13 mission of 1970 was technically a failure, since the astronauts were forced to abort their planned lunar landing, but many historians consider it to be among NASA's finest achievements. The astronauts reached home safely against almost impossible odds. And the ability to make accurate calculations quickly and efficiently proved critical to saving their lives.

People have always used counting aids, from fingers and toes and slide rules to handheld calculators and modern computers. In ancient times, they would perform calculations in much the same way we do with paper and pencil, except that they used counting stones or beads to represent a given quantity, and a specific position to represent value, instead of the abstract symbols we use today. These devices were known as abaci. One type of abacus is a specially marked flat surface used with counters, called a counting table. Ancient Greeks used a counting slab, a simple stone slab with incised parallel lines to mark the place values. Small stones, called calculi, were used with the tables in Greece and Rome; stamped metal coins were used in continental Europe and England. Similar devices were used for the next 2000 years all over the world: the English Exchequer didn't stop using the counting table for tallying tax payments until about 1826. Counting tables may have become obsolete, but the word "counter" has survived in our modern lexicon, to describe a flat surface upon which a business transaction can be conducted.

The second form of abacus is a frame with beads strung on wires. The Chinese are usually credited with inventing this particular form, which may be the world's earliest calculating machine. Use of a bead-frame abacus is first mentioned in historical records around AD 190, although the version we are most familiar with did not become common until about 1200. It usually has 13 vertical wires, with seven beads on each wire, housed in a rectangular frame. The device reached Japan in the seventeenth century, and from there spread to Russia. A slightly different form of the bead-frame abacus was found during archaeological excavation of an Aztec site in Central America. Dated to around AD 900, that abacus is made of corn kernels threaded on a string within a wooden frame, with seven beads in each of 13 columns.

Such counting devices had their uses, but in nineteenth-century England a mechanized alternative sprang from the prodigious imagination of a man named Charles Babbage. The son of a banker in London, Babbage was born to tinker: as a child he loved to take apart his toys to see how they worked. He taught himself algebra, and he was so well read in the mathematics of his day that when he entered Cambridge University's Trinity College, he found himself well in advance of his tutors on the subject. After finishing his studies, he was offered a prestigious professorship at Cambridge. Although Babbage found himself cut off from the family fortune when he married without his father's permission, he was still sufficiently well off to pursue his love of invention. Among other things, he invented a speedometer and a cowcatcher, a device that could be attached to steam locomotives to clear cattle from the tracks. His greatest passion was the design and construction of elaborate "thinking machines" capable of making rapid, accurate calculations without human intervention.

People seemed either to love Babbage or to hate him. His fans included Charles Darwin and Ada Lovelace, daughter of the Romantic poet Lord Byron, who grasped the significance of Babbage's "thinking machines" even as her contemporaries were ridiculing his obsession. Detractors included Ada's mother and the poet Thomas Carlyle, who once described Babbage as "a cross between a frog and a viper." True, Babbage could be tiresome, even pompous at times, and he had a gift for sly self-promotion, but when he chose to, he could also be quite charming. He loved numbers and found minute details endlessly fascinating, and he compiled a collection of "jest books" to scientifically analyze the "causes of wit." Imagine the comedian Jerry Seinfeld stopping after every joke to explain to his audience why it was funny. Babbage just couldn't help himself. At Cambridge he had cofounded the Analytical Society—essentially the first collegiate math club—and once, as a "diversion," he drew up a set of mortality tables, now a basic tool of the modern insurance industry. ("A man with such a head for numbers and flair for flattery was bound to end up in life insurance," the historian Benjamin Woolley quipped in *The Bride of Science*.) Babbage even cracked the supposedly unbreakable "Vigenere" cipher around 1854; this feat is considered by many historians to be the most significant breakthrough in cryptoanalysis since the ninth century. But these were mere fripperies; his true interest lay elsewhere.

Babbage first conceived of a calculating machine in 1821, when he was examining a set of mathematical tables with the astronomer John Herschel (the son of William Herschel). Such tables were used to make calculations for astronomy, engineering, and nautical navigation, but the table entries were calculated by hand and hence were riddled with errors. This meant that the resulting calculations were often wrong, no doubt causing any number of shipwrecks and engineering travesties. "I wish to God these calculations had been executed by steam!" Babbage exclaimed in exasperation after finding more than a thousand errors in one table. And he embarked on a lifelong quest to mechanize the process.

Exactly how this might be accomplished wasn't immediately clear, but during Babbage's many trips to Paris around the same time, he heard of a novel scheme used by the French mathematician Gaspard Riche (Baron de Prony). France had recently switched to the metric system of measurement. This gave scientists a much-needed standardized system to measure and compare results, but it also required a whole new set of calculating tables. The sheer number of calculations was beyond what could be accomplished by all the mathematicians in France, so Riche established calculating "factories" to manufacture logarithms the same way workers manufactured mercantile goods.

Each factory employed between 60 and 80 human "computers." But they weren't trained mathematicians; they were mostly out-of-work hairdressers who had found their skill at constructing elaborate pompadours for aristocrats much less in demand after so many former clients lost their heads at the height of the French Revolution. Riche had hit upon a rote system of compiling results based on a set of given values and formulas, and the workers just cranked out the answers in what must have been the world's first mathematical assembly line. Babbage figured that if an army of untrained hairdressers could make the calculations, so could a computing "engine." In fact, a mechanical calculator for adding and subtracting numbers, called the arithmometer, had just been invented.

Suitably inspired, Babbage returned to England and designed his first "difference engine," which created tables of values by finding the common difference between terms in a sequence. Powered by steam, it was limited only by the number of digits the machine had available. He built a demonstration model of his "thinking machine," one of the most complex machines ever built in the nineteenth century. The size of an upturned steamer trunk, it had two brass columns supporting two thick plates, and in between was a confusing array of gears and levers. Small wheels with numbers engraved in their rims displayed the results of the calculations. Babbage proudly displayed his prototype to the many visitors he entertained in his Dorset Street home, although most did not share his love of numbers.

In an effort to pique their interest, he put on a dog-and-pony show: he would announce that the machine would add numbers by two, then begin turning the crank. The wheels with the carved figures would begin displaying the predictable sequence: 0, 2, 4, 6, 8, and so on. The machine would do this for roughly 50 iterations. Just when the audience was becoming bored and restless, the number would suddenly leap to a new, seemingly random value and then continue the adding-by-two sequence from there. To Babbage's contemporaries this was nothing short of miraculous. In reality, Babbage had programmed the engine to perform one routine for a given number of turns of the crank, and then to jump to another program—he called this a "subroutine"—for another few turns before returning to the original routine.

Still, the difference engine was little more than a very large, complicated calculator. After 10 years of work, Babbage abandoned the project to work on a second version. It didn't help that he was a perfectionist who was never satisfied with his work, and continually revised his blueprints for the machine. He spent thousands of pounds of government funding to rebuild the same parts over and over. Small wonder that the British government decided to suspend funding for his work in 1832 and terminated the project altogether in the 1840s. So Babbage never built any of his designs, other than the small prototype machine in his home.

This is a shame, because the second version of the difference engine was the precursor to Babbage's design for the "analytical engine," now recognized as the forerunner of the modern computer. This new, improved machine wouldn't just calculate a specific set of tables; it would solve a variety of math problems by using the instructions it was given. The design was loosely based on the notion of a cotton mill, so there was a memory function called a "store," and a processing function called the "mill." As with the difference engine, numbers would be stored in columns of cogs, each cog representing a single digit. But to control the cogs, the analytical engine would use punched cards based on the weaving cards developed to "program" looms to weave particular patterns—essentially pre-Victorian floppy disks or CD-ROMs. When strung together, these cards would enable the machine to perform "loops," where a sequence of instructions would be repeated over and over, or "conditional branching," where one series of cards would be skipped and another read if certain conditions were met. In this way, the analytical engine was intended to have a rudimentary decision-making capability.

Nothing like the analytical engine had ever been seen before, and even other scientists and engineers found the concept difficult to grasp. Babbage published extensively on his work, giving rise to heated debate. Robert Peel, the head of England's Tory administration at the time, denounced the analytical engine as a "worthless device, its only conceivable use being to work out exactly how little benefit it would be to science." Over the years, Babbage would fill about 30 volumes with various designs and formulas for the analytical engine, but the machine was never built. It would have required tens of thousands of parts, intricately assembled into a frame the size of a small locomotive, with a similar amount of steam power needed to drive it. Some historians have speculated that Babbage's machines were never built because they demanded a level of engineering sophistication that simply didn't exist in pre-Victorian England. But it turns out that the difficulty was mostly a matter of money. In the 1990s, a team of scientists at London's Science Museum built a working model of the second difference engine, using only the materials and tools that would have been available in Babbage's day—and it worked. The machine is now prominently displayed in the museum.

In much the same way that the Apollo 13 mission snatched victory from the jaws of defeat, Babbage's analytical engine is deemed one of history's most successful failures. It would take revolutionary developments in electromagnetism, wireless radio, the invention of the transistor, and countless other technological innovations over the next 100 years before the first computers became available in the 1950s. It would be another 40 years after that before the first personal computers became commonplace. Babbage's thinking machine anticipated almost every aspect of its modern counterparts.

In 2002, the cyberpunk authors William Gibson and Bruce Sterling collaborated on a historical thriller, *The Difference Engine*, in which Babbage perfects his steam-powered analytical engine. They envision an alternative history in which the computer age arrives a century earlier, Lord Byron becomes prime minister instead of dying of syphilis at 32, and a rebellious group of subversive, antitechnology Luddites conspires to overthrow the intellectual elite in power. The authors confined their alternative vision to the nineteenth century. But thanks to the ability to process mind-boggling amounts of data quickly and accurately, modern computers have made vast breakthroughs in science possible: the mapping of the human genome and the discovery of the top quark are just two of the most prominent examples. We can only imagine how our daily lives, science, and the course of history might have been altered if Babbage's ingenious engines had given us comparable computing power in the nineteenth century.

10

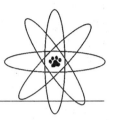

Emperors of the Air

People in the medieval English village of Malmesbury no doubt thought their local clergyman had gone mad when, in 1010, a humble Benedictine monk named Eilmer jumped from the 150-foot rooftop of his abbey, wearing a pair of crude wings he'd fashioned from willow wood and cloth. Eilmer managed to glide a good 600 feet, passing over the city wall before crash-landing in a small valley near the river Avon, and breaking both his legs. Malmesbury Abbey still boasts a stained-glass window in honor of Brother Eilmer, although a colorful local pub called The Flying Monk has been replaced by a modern shopping center.

Eilmer of Malmesbury was probably not the first person to attempt such a stunt; he certainly wasn't the last. A century later, a man in Constantinople fashioned a pair of wings from fabric and jumped from the top of a high tower. He wasn't as lucky as Eilmer, and died from the fall. In sixteenth-century France, an inventor named Denis Bolor tried to fly with wings that flapped using a spring mechanism. In Italy, Leonardo da Vinci designed (but never built) flying machines based on the wings of birds. And in 1678 a French locksmith tried to fly with wings modeled after the webbed feet of the duck. None of these attempts met with much success.

Ancient Greek philosophers like Aristotle identified air as one of four primary elements, along with earth, fire, and water. But air has more in common with water than might be immediately apparent. Air is a gas and water is a liquid, but in the realm of science both fall into the category of fluids. In physics, a material is considered a fluid if the amount of force needed to change its shape is determined by how quickly it changes. For a solid, the force needed to change its shape is determined by how much it changes. For example, it takes the same amount of force to break a twig quickly as to break it slowly. But moving your hand through a body of water quickly will deform the liquid more than moving your hand through it slowly. The same phenomenon happens with air, as anyone who has ever stuck a hand out the window of a fast-moving car can attest. This notion of air as a fluid lies at the heart of aerodynamic theory.

A skit on the quirky BBC radio program *People Like Us* featured a satirical "interview" with a fictional airline pilot, who is asked to explain how an airplane stays in the air. The interviewer admits that he studied the subject in physics class at school but has never been clear about how it worked, because there was so much theory. "Well, theory is why it stays up!" the pilot snaps brusquely. "If you took away the theory, even for a moment, it would just plummet like a stone!" But he does answer the question more or less correctly: "Basically it's all to do with pressure differences between the top and the bottom of the wing."

An Englishman, Sir George Cayley, in a paper published in 1810, was the first to articulate the principles of flight that govern winged aircraft. Aerodynamic theory has to do with two pairs of opposing forces: lift and weight (the pull of gravity), and thrust and drag. The symmetry is an example of Newton's third law of motion: for every action (lift, thrust) there is an equal and opposite reaction (weight, drag).

The most popular concept of aerodynamic lift is based on something called the "Bernoulli principle," named after the eighteenth-century Swiss physician who discovered it. Daniel Bernoulli came from a long line of mathematicians, but his father forced him to study medicine instead because it was a more practical profession. The English physician William Harvey—who was the first to observe that the human heart worked like a pump to force blood through the arteries so it could circulate through the body—encouraged Bernoulli to combine his love for mathematics with his medical training to discover the basic rules that govern the movement of fluids.

One day Bernoulli conducted a pivotal experiment: he punctured the wall of a pipe filled with fluid with a small, open-ended straw. He noticed that the fluid would rise up the straw, and the degree to which it would do so was directly related to the fluid's pressure in the pipe: the higher the pressure, the higher it would rise, much like a tire pressure gauge. His technique was soon adopted by physicians all over Europe, who used it to measure patients' blood pressure by sticking glass tubes into the arteries. Bernoulli went one step farther and applied this discovery to his earlier work on the conservation of energy. According to Newton's laws of motion, a moving body exchanges kinetic energy for potential energy as it gains height. A similar phenomenon occurs with a moving fluid: it exchanges its kinetic energy for pressure. This is the

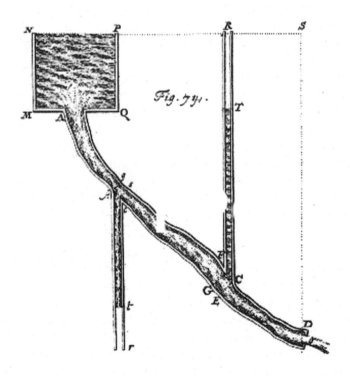

**Bernoulli's diagram
of pressurized fluid flow**

Bernoulli principle. It simply states that the pressure of any fluid decreases when the speed of the fluid increases. So high-speed flow is linked to low pressure and low-speed flow to high pressure.

An airplane's wings are designed to create an area of fast-flowing air (and hence low pressure) above the surface. It doesn't matter whether the object is moving through still air, or whether the air moves around the object. It's the relative difference in speeds between the two that creates lift. A wing is basically an airfoil, with a leading edge that is angled to "attack" the air in such a way that it increases the speed of the airflow above the wing, decreasing the pressure there. The aeronautics professor Scott Eberhardt prefers to think of the wing as a "scoop," diverting circulating air to the wing's angle of attack; he describes lift as the result of large amounts of air being pulled down from above the wing. The end result of both conceptual approaches is that the air pressure becomes greater underneath the wing than above, and that combination produces lift. When the lift becomes greater than the object's weight, the object will begin to rise.

The greatest limiting factor in flight is drag, best defined as the force exerted on an object as it moves through a fluid. We experience this as air or water resistance. Drag increases rapidly with flight speed. For example, running into a high wind requires more force to keep moving than running with the wind. The force that counters drag is known as thrust and is usually generated with jet engines and pro-pellers. To make flight possible, all four forces must maintain a con-tinually shifting delicate balance. When an airplane reaches its cruise altitude, the thrust is equal to the drag, and the lift is equal to the weight. If there is more drag than thrust, the plane will slow down. If the opposite is true, the plane will speed up. If the lift is less than the weight of the airplane, the plane will descend; whereas increasing the lift will cause the plane to climb to a higher altitude.

Maintaining the delicate balance between forces is easier said than done, as any participant in the "flugtag" competitions sponsored by Red Bull can attest. Would-be aviators—usually teenagers and college students, who have the leisure to pursue such exploits—build their own human-powered aircraft machines and then push them off a 30-foot platform (deliberately built over water) to see how far they can fly. The designs are often creative—there have been machines shaped like pigeons, giant cows, bowler hats, and Homer Simpson—but usually exhibit little in-depth understanding of the complexities of aerodynamic theory. Most of the machines drop like a stone into the water without ever achieving any semblance of flight. But then, that's half the fun for the competitors. Eilmer of Malmesbury would have heartily approved.

How do these principles apply to our wacky medieval monk? Eilmer generated an initial burst of thrust by leaping off the abbey roof. This action increased the flow of air and lowered the pressure on the upper part of his wings to create lift, enabling him to glide through the air. But the forces of drag and gravity kept increasing and eventually overcame the forces of thrust and lift. In mid-flight, the monk apparently tried to generate more thrust by frantically flapping his wings like a bird. This destabilized the delicate balance between the forces, and he "stalled out," falling to the ground. Stalling is what happens when the leading edge of the wing is tilted too sharply, pulling the airflow away from the upper surface, and causing the lift to vanish. It spells disaster for any flight. The German-born aviator Otto Lilienthal plummeted to his death in 1896 when his glider stalled out after being hit by a particularly strong gust of wind.

The earliest flying machines didn't have wings. In 1780, two French paper makers, the Mongolfier brothers, noticed that smoke from a fire built under a paper bag would cause it to rise into the air. The hot air inside expanded, and thus weighed less, by volume, than the surrounding air. The Mongolfiers put the concept to practical use in 1782 when they built the first hot-air balloons. The first passengers were a duck, a sheep, and a rooster, but within a year the brothers became the first humans to experience free flight when they floated above Paris for half an hour at an impressive 9600 feet. Ballooning soon became a popular recreation, and in 1785 John Blanchard made the first aerial crossing of the English Channel in a hot-air balloon.

The most serious drawback to balloons was that they were difficult to steer, and therefore subject to the whims of the wind. In the mid-nineteenth century, another Frenchman, Henri Giffard, built the very first airship, or dirigible. Its enormous envelope was shaped like a cigar and filled with hydrogen, a gas that is lighter than air at normal temperatures. Giffard's machine was also outfitted with a steam engine, which turned a propeller to create thrust, and rudders to help with steering. He made his first successful flight on September 22, 1852, traveling roughly 17 miles at 6 miles per hour.

Dirigibles remained a novelty for nearly 50 years, until a wealthy Brazilian emigrant named Alberto Santos-Dumont became absorbed with building them. Further incentive came in 1901, when the French Aero Club offered a prize of 100,000 francs to the first airship to complete a journey from the club's airstrip at Saint-Cloud to the Eiffel Tower and back in less than 30 minutes. On his second attempt, Santos-Dumont completed the journey in 40 minutes—not good enough to earn the prize. On his third attempt, he was forced down by a hydrogen leak and crash-landed into a local restaurant. The airship's envelope was ripped to shreds, but the framework dangled from what remained of the building's walls just long enough for Santos-Dumont to climb down to safety. The Brazilian quickly constructed a replacement dirigible, and on October 19 he completed the trip a mere 40 seconds past the time limit. The judges grudgingly awarded him the prize money anyway, and he donated it to Parisian charities. Santos-Dumont became a familiar, dapper figure around town, occasionally bar-hopping in a little dirigible that he moored to nearby lampposts.

Early dirigibles could only fly in calm or nearly calm weather. In strong winds, they would fly in slow circles. At the dawn of the twentieth century, lightweight engines were invented that could generate sufficient power to overcome strong winds, and dirigibles became truly practical. This was the era of the giant zeppelins, named after Germany's Count Ferdinand von Zeppelin. His airships were used in World War I to drop bombs over London, and as commercial passenger carriers. Dirigibles still had major drawbacks. They could break up in midair if the altitude shifts became too severe. That was the fate of the first U.S. airship, the *Shenan-doah* which got caught in a powerful updraft during its maiden voyage in 1925. And dirigibles could explode from the combustible gases inside the envelopes. The disastrous explosion of the *Hindenburg* in 1937 killed 36 of the 97 passengers onboard. That tragedy effectively ended the use of dirigibles as commercial passenger carriers. A few blimps are still in use today, but they are no longer filled with hydrogen.

Clearly an alternative flying machine was needed, with both a suitable source of power to produce and maintain thrust and a means of steering the aircraft. Two American brothers, Wilbur and Orville Wright, began experimenting with gliders in 1900. They noticed that when a bird flew, it twisted the back edge of one wing upward and the back edge of the other downward to change direction—a phenomenon the Wrights called "warping." They incorporated that concept into their glider design by attaching cords to the wingtips. Pulling on the cords would warp the shape of the wings slightly, just enough to enable them to steer the craft. They first flew such a glider in 1902 at Kitty Hawk, North Carolina, and then returned the following year with their newest machine. The "flyer" followed the same basic design as the glider, with the addition of an engine and propellers to generate sufficient thrust to overcome drag and keep the machine in the air. They made their historic first flight on December 17, 1903, traveling 120 feet in 12 seconds, even though they barely got above 10 feet in the air.

In Europe, no one was aware of this achievement, since the Wrights didn't demonstrate their machines publicly until 1908. Santos-Dumont had by then abandoned dirigibles for winged aircraft. He succeeded in flying such a craft on November 12, 1906, and his *demoiselle* ("dragonfly") design, created in 1909, became the forerunner of the modern light plane. That same year, Louis Bleriot flew his monoplane across the English Channel from France to England; and in 1910, when France began issuing pilot's licenses, Elise Deroche became the first licensed female pilot. In America, Alexander Graham Bell, who dreamed of building flying machines with telephones attached, collaborated with other inventors to construct the *Silver Dart*, which flew for a full ½ mile. At long last, mankind had mastered the air.

The psychology of dream interpretation is an inexact science at best—a physicist might argue that it isn't a science at all. But more than a third of human beings are believed to have had at least one flying dream, an experience that can be traced back to the Babylonians and Egyptians. In fact, every ancient culture seems to incorporate some notion of human flight in its legends and mythology. The Greeks had Icarus, whose father, Daedalus, fashioned wings of feathers for them both to escape their island exile—with fatal results for Icarus. In South American mythology, humans use artificial feathered wings to leap from tall towers; and China's Emperor Shun was said to have escaped from a burning granary using a crude pair of wings. Around 3500 BC, a Babylonian king, Etena, was pictured on a coin riding on an eagle's back. In Africa, a Ugandan king, Nakivingi, was believed to employ an invisible flying warrior named Kibaga, who hurled rocks from the air onto the king's enemies. Mojave tribes worshiped Mastamho, a warrior who could transform himself into an eagle. The founder of the Incan people, Auca, supposedly had wings and could fly. And ancient Indian Vedic writings mention both magical and mechanical flying machines, called *vimanas*.

On a purely physical level, flying dreams are related to the part of the central nervous system that regulates equilibrium; wearing a blood pressure cuff to bed, or being rocked in a hammock, will induce dreams of flying, according to the findings of laboratory studies. But Carl Jung believed that dreams of flying signified the human subconscious expressing a desire to break free of limitations—to rise above the difficulties and petty concerns that plague our daily lives. That might have been what Eilmer of Malmesbury felt when he made his historic leap from the abbey. For one brief moment, he was free.

11

Disorderly Conduct

June 1871
Maxwell and his demon

Lyra, the winsomely headstrong heroine of Philip Pullman's fantasy novel *The Golden Compass*, inhabits a world where people's souls take visible form as totemic animal-shaped "daemons," and physics falls under the rubric of experimental theology. Among the more intriguing creatures she encounters in her journeys is a murderous mechanical device, shaped like a beetle. Inside it is a clockwork mechanism with a "bad spirit" pinned to the spring by a spell through its heart. The thing is single-minded in its mission to seek out and destroy its target, and because the clockwork gets its energy from the spirit, it will never run down until it fulfills its mission. In fact, it gets stronger and stronger, feeding off the hatred that fuels it.

Hatred might work that way, at least metaphorically in science fiction, but here in Earth's mortal coil, it is an accepted maxim that nothing—especially energy—can go on forever. So it was something of a surprise when, in 1618, a London physician and alchemist named Robert Fludd made an extraordinary claim: he thought he had discovered a means of producing enough energy to operate a waterwheel to grind flour in a mill, without depending on nature to provide the powering stream. Waterwheels date as far back as 20 BC and the days of the Roman Empire, and the basic concept is still used today in hydro-electric power stations.

Fludd's idea was to use the waterwheel to drive a pump, not just to grind flour. That way, the water that turned the wheel could then be pumped back up into a standing reservoir of water, which could be used to run the mill indefinitely. It seems a reasonable enough claim, on the surface. But Fludd's device is an early example of a perpetual motion machine, a hypothetical device capable of operating forever on a fixed supply of energy. Unfortunately, such schemes violate the basic laws of thermodynamics, which dictate that mechanical systems cannot operate continuously without some form of external assistance. The first law of thermodynamics is that energy is conserved. All this means is that you can't get something from nothing; energy is never "free." The second law is that any machine, no matter how well built, will gradually slow down and stop if no extra energy is added. Robert Park, a physicist at the University of Maryland, sums it up best in his book *Voodoo Science:* "The first law says you can't win; the second law says you can't even break even." Physicists can be such killjoys.

Fludd's idea is a case in point. For his device to work, the water would have to be lifted back up the same distance it fell while turning the wheel, working against gravity. As anyone who's ever toted a gallon of water on a day hike can attest, water is a heavy substance. Pumping all that weight back up into the supply tank requires energy—so much that all the energy generated in turning the wheel would be needed just to raise the water back to the reservoir. There wouldn't be any left-over energy to grind the flour. In fact, even without grinding flour, Fludd's wheel couldn't keep running forever. Energy would inevitably be lost through the heat and sound generated by friction in the machinery. A bicycle is one of the most energy-efficient machines ever invented, but its wheels won't turn indefinitely unless its rider keeps pedaling. Friction between the tires and the asphalt, not to mention aerodynamic drag, conspire to slow the bicycle down. It takes extra energy to overcome those opposing forces and maintain forward momentum.

This phenomenon is known among physicists as entropy. It's a tricky, almost counterintuitive concept that has flummoxed many a student of introductory physics—not to mention would-be inventors like Fludd—but it is the reason perpetual motion machines cannot exist. It is best described as a measure of how much disorder there is in a given system. Why is this important? The more order is present in a system, the more energy is available for work. The greater the disorder, the less energy is available for work—which might explain why some people have a hard time getting anything done in a messy environment.

Picture a set of dominos, neatly arranged on end in a long line; they are very well ordered, and thus have the potential to produce considerable energy. A mere flick of the finger could release that potential energy and send the dominos toppling one after the other. But in the process, they would become disordered and lose energy; there is no way to restore the order without bringing in additional energy, in the form of a human being to line up the dominos again. The same is true of Fludd's waterwheel. M. C. Escher produced a lovely drawing of a perpetual waterfall, but like many of his sketches, it's an optical illusion achieved by playing with geometric perspective. Real fountains need some sort of external energy source, such as electricity or solar energy, to keep running. In an ideal world, energy could be easily converted from one type to another: for example, heat energy could be converted into mechanical energy to rotate Fludd's waterwheel (or run Escher's waterfall), and back again, without losing any energy to entropy. But we don't live in an ideal world; in the natural world, everything tends toward disorder.

That hasn't kept hopeful inventors from trying for centuries to flout the laws of thermodynamics. A fifteenth-century Italian physician and alchemist claimed to have invented a self-blowing windmill, and in the 1670s the bishop of Chester designed a number of supposed perpetual motion machines. The indefatigable clockmaker Johan Ernst Elias Bessler designed some 300 perpetual motion machines in the early eighteenth century, and actually succeeded in building a wheel that rotated for 40 days in a locked room in a neighboring castle. No one was allowed to study the machine closely—that fact is suspicious in itself—but historians surmise that Bessler secreted a clockwork mechanism in the large axle of the wheel to keep the device running so long. Even today, the U.S. Patent Office is inundated each year with applications for patents on perpetual motion machines and their first cousins, so-called "free energy" schemes that claim to create energy out of nothing. Most of these are rejected automatically on the basis of thermodynamics.

The roots of thermodynamics lie in the work of a little-known French physicist named Sadi Carnot. His relative obscurity today is a bit surprising, since he was the son of one of the most powerful men in France during Napoleon's rule. Like many sons of the aristocracy, Carnot was well educated, choosing to study engineering after a brief and unsatisfying stint in the military. After Napoleon's defeat and his father's subsequent exile, Carnot became interested in steam engines; he was convinced that France's failure to improve this very basic technology had contributed to Napoleon's downfall at the hands of the British, whose steam engines were much more efficient. But his quirky obsession yielded some intriguing ideas about how energy is produced, forming the basis for his only major published work, *Reflections on the Motive Power of Fire*, which appeared in 1824. In it, Carnot described a theoretical heat engine that produced an amount of work equal to the heat energy put into the system.

Carnot was not a proponent of perpetual motion; he conceded that a perpetual motion machine would be impossible to construct, since it would lose at least a small amount of energy to friction, noise, and vibration. But his efforts helped establish the total amount of useful work that a heat engine, operating between two temperatures—a hot energy reservoir and a cold energy reservoir—could produce. This is known as the "Carnot cycle." Modern-day refrigerators are based on a similar concept, drawing energy from differences in temperature. A gas (usually ammonia) is pressurized in a chamber, and the pressure causes it to heat up. The coils on the back of a refrigerator serve to dissipate this heat, and the gas condenses into an ammonia liquid, still highly pressurized. The liquid then flows through a small hole to a second, low-pressure chamber. The change in pressure causes the liquid ammonia to boil and vaporize, dropping its temperature to -27 degrees Fahrenheit and making the inside of the refrigerator cold. The cold ammonia gas is then sucked back into the first chamber, and the entire cycle repeats. A refrigerator needs a continual influx of electricity to continue operating, of course, so it is not a perpetual motion machine.

Carnot's work went largely unnoticed until after his death from cholera at the age of 32. It would fall to Rudolf Clausius and William Thomson (Lord Kelvin) to extend his work, making it the foundation for modern thermodynamics, in the 1840s and 1850s.

Unlike the laws we make in society, physical laws are absolute. We can't just vote to suspend gravity; the same goes for thermodynamics.

But though the laws of physics can't be broken, sometimes they can be bent. The most famous evasion of thermodynamics was proposed in the nineteenth century by the Scottish physicist James Clerk Maxwell, who was born in Edinburgh. Maxwell displayed a curiosity about how things worked even while still a toddler. As an adult, he earned a degree in mathematics from Cambridge University in 1854. It was during his six-year tenure as chair of natural philosophy at King's College in London that Maxwell conducted his most important work on electro-magnetism. But he was equally intrigued by thermodynamics.

According to the second law of thermodynamics, heat cannot flow from a colder to a hotter body, any more than water can flow uphill. Maxwell rather mischievously proposed that gas molecules at higher temperatures merely have a high *probability* of moving toward regions of lower temperature. The laws could, in theory, be outwitted. To illustrate this point, Maxwell conceived of an ingenious thought experiment in 1871. He envisioned an imaginary microscopic crea-ture able to wring order out of disorder to produce energy, chiefly by making heat flow from a cold substance to a hot one. This little imp guards a small pinhole in a wall separating two compartments of a container filled with gas, similar to Carnot's hot and cold reservoirs and the high-pressure and low-pressure chambers used in refriger-ators. The creature controls a shutter that covers the hole, which he can open or close whenever he wishes.

The gas molecules in both compartments are highly disordered, randomly moving in all directions, so there is very little energy avail-able for work. The molecules' average speed and temperature are all the same in each compartment, at least at the outset. But this state of equi-librium doesn't last. Soon some molecules start moving faster than the average speed and hence get hotter, while others start moving more slowly and hence become cooler. Normally a closed system will correct such imbalances; heat will be transferred from the hotter to the colder molecules until they are all once again exactly the same temperature.

In Maxwell's thought experiment, whenever the demon sees a fast molecule in the right-hand compartment approach the pin-hole, he opens the shutter briefly to let it through to the left side, and also lets slower molecules from the left side pass to the right. Over time, the average speed of molecules on the left increases, while that in the right compartment decreases. So the gas on the left gets steadily hotter and that on the right gets colder. The system is becoming more ordered, and so there is more potential energy available for work. The creature is creating a temperature difference between the gases in the two compartments, which can be harnessed to make a machine do physical work.

This is "Maxwell's demon," so named because it presents a paradox that has stumped many scientists since it was first proposed. It does indeed appear to violate the second law of thermodynamics and overcome entropy. But before you conclude that thermodynamics is a crock, remember that we don't live in Maxwell's theoretical universe. The "demon" is essentially a clever illusion, and Maxwell himself supplied two reasons why the demon couldn't exist in the physical world. The first is simply that the laws of thermodynamics apply to molecules en masse, not as individuals. It is statistically impossible to sort and separate individual molecules by speed or temperature, any more than one could throw a glass of water into the sea and get back the exact same glass of water, right down to the last molecule.

As for the second reason, Maxwell said that we could, in theory, break the second law, but to do so we would have to know the exact positions and speeds of each and every molecule. Even assuming that we could surmount statistical improbability to manage this feat, the very act of gaining the information would require so much energy that we would use up all the work output of any machine we could drive with such a system. So Maxwell's imaginary sorting demon itself requires energy to operate, playing much the same role as the "bad spirit" that fuels Pullman's perpetual clockwork beetle.

12

Queens of Science

November 1872
Death of Mary Somerville

In David Auburn's Pulitzer Prize-winning play *Proof*, a young woman in Chicago, named Catherine, loses her father, a brilliant but mentally unstable mathematician, on the eve of her twenty-fifth birthday. One of her father's former students, Hal, is rooting through the 103 notebooks the father left behind, almost all of them filled with garbled nonsense, a testament to the severity of the dead man's mental illness.

Romance inevitably blooms between Catherine and Hal, and she makes a critical decision: she hands him a key to a locked drawer in her father's desk. There, he finds revolutionary mathematical proof of a problem that most mathematicians believed couldn't be solved. Hal concludes that the proof was done by her father, who had revolutionized the field twice before he turned 22. Somehow, in the midst of his madness, he managed to complete the most important work of his life. When Catherine reveals that she, not her father, wrote the proof, Hal is stunned. More significantly, he doesn't believe her.

Hal's reaction epitomizes a fundamental prejudice against women in math and science that dates back thousands of years and still pervades popular culture. At a math tournament in the teen comedy film *Mean Girls* (2004), each team must pick the weakest member from the other team to compete in a tie-breaking "sudden death" round. The boys don't even need to mull it over: they automatically pick the token female on the opposing team. Hal is a bit more enlightened, but the whiff of prejudice is still there. Because Catherine took only a few courses at Northwestern before dropping out to care for her father, Hal doesn't consider her a "professional" mathematician. He doesn't think she's capable of producing such advanced proof. It never occurs to him that she might have inherited her father's prodigious gift for numbers, even when Catherine tartly reminds him: "My education wasn't at Northwestern. It was living in this house for 25 years." Under her father's tutelage, she learned what a prime number is before she could read.

Around 600 BC, in ancient Greece, women openly studied math, philosophy, astronomy, acoustics, and medicine in the academies of Plato and Pythagoras; they wouldn't have equivalent opportunities for education again until the twentieth century. The last exceptional woman scientist of antiquity was Hypatia, the daughter of a professor of math and astronomy. She lived around AD 370 in Alexandria, Egypt, and developed her own instruments for astronomy and navigation and wrote several scholarly treatises while heading Alexandria's Neoplatonist school of philosophy. But her peaceful existence ended abruptly when Alexandria became part of the Christianized Roman Empire, with a corresponding distrust of the pagan philosophies she espoused. Hypatia was summarily tortured to death.

Such practices understandably had a dampening effect on enthusiasm for acquiring knowledge, and not just for women. Centers of learning in western Europe declined sharply after 200 BC, and the region entered the Dark Ages. By contrast, Asia was unusually enlightened during this period: in seventh-century Korea, for instance, Queen Sonduk was one of three female rulers of the Silla kingdom. She built the "Tower of the Moon and Stars"—the first astronomical observatory in the Far East. But for the most part, notable women in math and science were few and far between. During the Middle Ages, learning was strictly controlled by the church, and even members of the upper classes tended to be uneducated unless they joined the clergy.

The rise of the merchant classes in the early Renaissance produced an atmosphere more conducive to learning, aided by the invention of the printing press. In England, the ascension of Elizabeth I encouraged noblemen to educate their daughters to be fit companions for the highly intelligent queen. Among them was the countess of Pembroke,

Mary Sidney Herbert, a chemist of some repute, although her literary works have overshadowed her scientific accomplishments. Still, this period proved to be only a brief respite. Elizabeth's successor, James I, forbade his daughter to be taught, declaring, "To make women learned and foxes tame has the same effect—to make them more cunning."

Much of Western society seemed to share James's misogynistic views. Learning was deemed "unnatural" for women. There were even men who claimed that women who used their brains would be unable to bear children, since brain work would draw blood away from the womb and cause it to shrivel up. Such superstition was exacerbated by the witchcraft hysteria that swept through Europe in the mid-seventeenth century. Like physicians, midwives, and alchemists, academically inclined women fell under suspicion. A prominent French female mineralogist named Martine de Bertereau du Chatlet was accused of witchcraft and executed in 1642.

The tide began to turn in the eighteenth century with the rise of salon culture. Aristocratic women were far more likely to receive a rudimentary education, although excessive learning was still frowned on. Consider the case of Émilie du Chatelet. The daughter of a French aristocrat, Émilie was a very precocious child: she had read Cicero and studied math by age 10, and she was fluent in several languages by age 12. Her unquenchable thirst for knowledge greatly irked her father, who complained to her uncle, "I argued with her in vain; she would not understand that no great lord will marry a woman who is seen reading every day." Her mother voiced similar concerns, declaring that there was no proper future for a daughter who "flaunts her mind and frightens away the suitors her other excesses have not driven off." They were wrong: at age 19, she married the 34-year-old marquis du Chatelet.

Marriage did nothing to dampen her "excesses," and not all of her interests were scholarly; she was remarkably unconventional for her day. She became the lover of the duc de Richelieu, who hired professors of math and physics to give her a more formal education. In 1733, she donned men's breeches and hose and gatecrashed the Café Gradot, a Parisian coffeehouse where leading scientists, mathematicians, and philosophers often gathered. It was there that she met Voltaire, with whom she had a 10-year affair. Voltaire paid for renovations to her husband's country estate (the marquis must have been either very cosmopolitan or extremely naive), and the two lived and worked there in separate studies, amassing a library of 10,000 volumes. This is where Émilie du Chatelet completed her translation of Isaac Newton's *Principia*, hanging pipes, rods, and wooden balls from the rafters to duplicate Newton's experiments. She is best known for her three-volume treatise on the German mathematician Leibniz, and for collaborating with Voltaire on his own treatise on Newton. But her unconventionality didn't save her from meeting the same fate as many women of her time: she died in her early forties from complications following childbirth.

Women who lacked Émilie du Chatelet's exalted position found society to be far less indulgent. In *Proof*, Catherine tells Hal about her childhood hero, Sophie Germain, who had no formal education. Germain discovered math as a child while browsing in her father's library, where she came across an account of the death of the Greek mathematician Archimedes. Legend has it that when the Roman army invaded his home city of Syracuse, Archimedes was so engrossed in studying a geometric figure that he failed to respond when a soldier questioned him. His distraction cost him his life, as the soldier summarily speared him to death. The young Sophie concluded that if someone could be so consumed by a geometric problem, geometry must be the most fascinating subject in the world, and she set out to learn about it.

She taught herself the basics of number theory and calculus without a tutor, and studied Newton and Leonhard Euler at night. Her parents did their best to dissuade her, confiscating her candles and clothing and keeping her bedroom unheated. Germain countered by hoarding candles and wrapping herself in bedclothes to keep warm while she studied in secret, even though it was sometimes so cold that the ink froze in the well. In 1794, when she was 18, the École Polytechnique opened in Paris to train future mathematicians and scientists, but it was open only to men. So she assumed the identity of a former student, Antoine August LeBlanc. Every week she would submit answers to problems under her new pseudonym. The course supervisor, Joseph Lagrange, soon noticed the marked improvement in a student who had previously shown abysmal math skills, and requested a meeting. She was forced to reveal her true identity, but far from being appalled, Lagrange was impressed and became her mentor.

In her twenties, Germain gained another mentor, the German mathematician Carl Friedrich Gauss. With his encouragement, she developed what are known as Germain primes: if you double a Germain prime and add 1, you get another prime number. For example, 2 is a prime number. Double that is 4, plus 1 is 5, and 5 is also prime.

Despite her success, Germain felt that the scientific community as a whole did not show her the respect she believed she'd earned. She had a point. One of her rivals, Siméon Poisson, was technically her colleague, yet he pointedly avoided having serious discussions with her, and openly snubbed her in public. Late in her life, Gauss arranged to have the University of Göttingen award Germain an honorary doctorate, but she died of breast cancer before the honor could be bestowed. Her death certificate identified her not as a mathematician but as "a single woman with no profession."

Germain had a contemporary counterpart in Great Britain. Mary Somerville was the daughter of a vice admiral in the British navy. She had only one year of formal schooling at a boarding school for girls, which she loathed, although she did acquire a taste for reading. While flipping through a women's magazine, she happened upon some mysterious symbols in the puzzle section and learned that they were algebraic expressions. Since it was unthinkable in nineteenth-century Great Britain for a young woman to purchase books at all, especially on such "unfeminine" subjects as math and science, she persuaded her brother's math tutor to buy them for her, and began reading them voraciously.

Like Germain, she fought her family's objections, studying at night by candlelight, and causing her father to lament, "We shall have young Mary in a straitjacket one of these days." When he took away her candles, she memorized texts during the day and worked out problems in her head at night. He was greatly relieved when she married her second cousin, Samuel Grieg, in 1804, and quickly bore him two children. But Grieg, a member of the Russian navy, had a low opinion of intellectual women, and the marriage was not a happy one. He graciously died three years later, leaving her with a comfortable inheritance. Now financially independent, she was free to study as she pleased, and she gradually compiled a small library of works that gave her a solid background in mathematics. In 1812 she married William Somerville, a surgeon in the British navy who supported her academic inclinations.

Mary Somerville was chiefly known for popularizing scientific treatises by Laplace and Newton for general readers. Her books earned her election to the Royal Astronomical Society, along with Caroline Herschel, sister of the astronomer William Herschel, who made a name for herself studying the stars. They were the first women to be so honored. Somerville was revising a mathematics paper the day before she died at age 92. Obituaries in London hailed her as the "queen of science."

Yet Somerville was still a creature of her era. She was humble, self-effacing, and the quintessential Victorian wife and mother. She encouraged her protégée, Ada Lovelace, to work at sewing as well at sums, to prove that "a mathematician can do other things besides studying x's and y's." Deep down, she believed that women were not as gifted as men when it came to scientific creativity, and doubted her own abilities. "I have perseverance and intelligence, but no genius," she once wrote. (Auburn's Catherine voices similar doubts, to such an extent that her father, in a rare lucid moment, upbraids her for underestimating herself.)

Lovelace was a different creature altogether; she considered herself a "poetical scientist," one who united reason with imagination. She was the daughter of the Romantic poet Lord Byron, whom she never knew, and as a girl she was actually encouraged to study math and science, in hopes that they would counter the romantic "excesses" everyone feared she had inherited from her notoriously unstable and wayward father. Dubbed the "enchantress of numbers," Lovelace is best known for her work with Charles Babbage and his "thinking machines," but she was also interested in what are known today as artificial intelligence and neural networks. She sought a "calculus of the nervous system," that would show mathematically "how the brain gives rise to thought and nerves give rise to feelings." But Lovelace was still a Byron at heart: in 1851 she applied her math skills to an elaborately ill-conceived gambling scheme, and lost a great deal of money betting on horses. She died of ovarian cancer in her thirties.

Today, education is no longer the exclusive domain of men, and women in math and the physical sciences have made considerable gains. The number of bachelor's degrees awarded to women in physics—the most mathematically intensive science—has risen from less than 8 per cent in 1975 to about 22 per cent in 2002. This is still in sharp contrast to biology, in which women earned more than 30 per cent of the bachelor's degrees awarded in 1966, and more than 60 per cent of those awarded in 2001. The number of PhDs earned by women in physics has climbed from less than 4 per cent in 1975 to roughly 15 per cent in 2002.

The numbers may have improved, but the negative stereotypes still linger. Cady, the heroine of *Mean Girls*, is urged by her peers not to join the high school math team, despite the fact that she's quite gifted: "It's social suicide!" she is told. Home-schooled in Africa for most of her early life (her parents are anthropologists), Cady didn't absorb the subliminal message that women can't—and shouldn't—do math and science. It doesn't even occur to her that this might stigmatize her as a "geek." High school peer pressure being what it is, it doesn't take her long to learn the lesson. Within weeks, she is pretending to be bad at math so as not to intimidate a cute boy in her calculus class.

Proof ends on a positive note, with Hal finally choosing to believe Catherine's claim. And Cady not only embraces her gift for calculus but also gets to be homecoming queen. She proves that a girl can be pretty, popular, and very smart, and even date the cutest boy in school, just by being true to herself. It's a message more young girls need to hear.

13

Calling Mister Watson

March 10, 1876
First documented transmission of human speech

One particularly engaging character in Neil Gaiman and Terry Pratchett's comic fantasy novel *Good Omens* (1990) is A. J. Crawley. Crawley is a lapsed demon who has incurred the wrath of his superiors in hell by attempting to thwart rather than bring about Armageddon—mostly because he's loath to trade in the sleek luxuries of life on Earth for hell's rigid austerity. His superiors send Hastur, a duke of hell, to bring him back to Pandemonium for punishment.

Crawley is able to escape his infernal adversary by disappearing down the wire of the telephone line; Hastur follows suit. An atomic-scale chase ensues at fiber-optic speeds through several exchanges. Crawley finally emerges from the line moments before an answering machine picks up, trapping his pursuer in the machine. Crawley has a narrow escape; the telemarketer who calls a few hours later and releases his pursuer from the machine isn't so lucky, and meets with a gruesome death. A cranky demon is a homicidal demon, especially after being cooped up in an answering machine for hours.

In the novel's fictional world, the size, shape, and composition of whatever physical form demons choose to take are not subject to the same laws of physics that keep mere mortals from accomplishing similar feats. So Crawley is easily able to transform himself into a bit of electric current that runs through the telephone wire. Apart from a handy plot device, his inventive escape route provides an apt analogy for what happens to sound waves whenever we make a telephone call.

Sound occurs when a vibrating object, such as a ringing bell, sends a pressure wave through the atmosphere. This moves the air particles around it, which in turn move the air particles around them, spreading the vibration through the air in a wave that ripples outward. Dropping a pebble into a calm body of water has a similar effect. Just like light waves, sound waves have their own spectrum of frequencies, known as pitch. A higher wave frequency means that the air pressure is vibrating faster per second and the ear hears this as a higher pitch. With lower frequencies, there are fewer vibrations per second, and the pitch is lower. But unlike light, sound can't travel in a vacuum. Objects can produce sound only when they vibrate in matter. Any scene from *Star Trek* depicting the sounds of explosions in deep space is wrong, but the tagline for the first *Alien* movie is correct: "In space, no one can hear you scream."

Sound waves are mechanical energy, whereas light waves are electromagnetic energy. It is precisely because of their mechanical nature that sound waves can do what Crawley did, and convert themselves into an electrical transmission. We saw in Chapter 8 that passing a coil of wire through a magnetic field will generate electricity within the coil, and passing electricity through a coil of wire will give rise to a magnetic field. A telephone's transmitter contains both a wire coil and a small magnet. Speaking into the transmitter causes the coil to vibrate in response to the sound waves within a magnetic field. This turns the sound wave into an electrical signal, which can be transmitted over the telephone wire. That current is detected by the receiver's coil, producing a second magnetic field. And this causes a thin membrane, similar to the human eardrum, to vibrate in response to the electrical signal, turning it back into sound.

The telephone has its roots in an earlier device, the electric telegraph. The notion dates back to 1763, when an anonymous letter published in a Scottish magazine discussed how a message might be sent across distances using 26 cables connecting the sender and receiver— a cable for each letter of the alphabet. Pulses of electricity could be sent along wires corresponding to the appropriate letters, and the receiver would pick up the corresponding electrical current emerging from each wire to read the message.

Unfortunately, the idea was too far ahead of its time. The technology did not yet exist to construct such a system: most notably, a steady source of electrical current, and a sufficiently sensitive detection method for the electrical signals. But with the subsequent inventions of the battery and the electromagnet, the British physicists William F. Cooke and Charles Wheatstone were able to develop a rudimentary telegraph in 1837. A metal key—a simple electrical switch with a pair of contact wires—is pressed up and down, making or breaking the electric circuit and transmitting the signal as a series of electric pulses, according to a predetermined code. Within two years, their system was being used to send messages between local railway stations, as much as 29 kilometers apart. British police relied on the telegraph in 1845 to help capture a fugitive murderer, John Tawell. Accused of killing his mistress in a village called Slough, Tawell escaped police custody and jumped onto a London-bound train. The local police telegraphed Tawell's description to London, and he was arrested as soon as he arrived.

The Cooke–Wheatstone approach was ultimately superseded by the telegraph system created by an American portrait artist turned inventor

The Cooke–Wheatstone telegraph

named Samuel Finley Breese Morse. He was born in Massachusetts in 1799 to a local pastor and eminent geographer, graduated from Yale University, and set out for England to study art. It was in 1832, on his way back to the United States after a three-year stint studying artistic styles in Europe, that Morse first heard about electromagnetism and its potential for telegraphic communication. He immediately set out to construct his own telegraph. His version sent pulses of electric current to an electro-magnet, which moved a marker to write marks on a strip of paper. Morse also invented his own system of dots and dashes, which could be combined to represent letters of the alphabet, called Morse code.

Morse persuaded Congress to approve $30,000 in funding to con-struct an experimental telegraph line from Washington, D.C., to Baltimore. And on May 24, 1844, he sent the message "What hath God wrought?" from the old Supreme Court chamber in Washington to his partner, Alfred Vail, in Baltimore. This officially opened the com-pleted line. Within four years, another 5000 miles of landline had been installed, and orders began to pour in from Europe for similar sys-tems, making Morse a very wealthy man. By 1869, the east and west coasts were connected by telegraph. The Associated Press began operation the following year, transmitting news throughout the world. By the early twentieth century, crowds would gather outside newspaper offices in major cities to hear election results and the scores of baseball games. The telegraph also played a pivotal role in military operations. It was used first at Varna during the Crimean War in 1854, and later in the American Civil War. Abraham Lincoln was the first U.S. president to be able to direct his armies in the field from the White House. Before the telegraph, presidents would wait days and sometimes weeks for news from the battlefields of distant wars.

As marvelous as the telegraph proved to be, it was capable of send-ing and receiving only one message at a time. Telegraph companies dreamed of a device capable of sending multiple messages simultan-eously over the same wire. Among those competing to be the first to succeed were Thomas Edison, Elisha Gray, and a budding young inventor named Alexander Graham Bell, who persuaded his future father-in-law to bankroll his efforts. The race to develop a multiple telegraph would lead to a rival technology that would ultimately make the telegraph obsolete: the telephone.

Bell was born into a family enthralled by the idea of sound and its possibilities. His grandfather was an eminent elocutionist, and his father developed the first international phonetic alphabet. This was a form of "visual speech" that deaf people could use to make specific sounds. One of young Aleck's first memories was of sitting in a wheat field trying to hear the wheat grow. His mother, Eliza, was almost totally deaf, but he soon discovered that he could press his lips against his mother's forehead and make the bones resonate to his voice. He also participated in his father's demonstrations of "visual speech": Aleck would be placed out of earshot and his father would ask a member of the audience to make a sound, which he would write down in his phonetic alphabet. Then Aleck would enter and imitate the sound, using his father's notation.

A gifted pianist, Bell learned early on to discriminate pitch, and by the time he was a teenager, he realized that a chord struck on one piano would be echoed by a piano in another room. He concluded that whole chords could be transmitted through the air, vibrating at the other end at exactly the same pitch. The effect is known as "sympathetic resonance." Objects like a piano's strings vibrate at very specific frequencies. If there is another object nearby that is sensitive to the same frequency, it will absorb the vibrations (sound waves) emanated from the other object and start to vibrate in response. Bell also knew that human speech is composed of many complex sound vibrations. In fact, while on vacation in Ontario one summer, he constructed an "ear phonoautograph" from a stalk of hay and a dead man's ear. (He acquired the ear from a local medical school.) Gruesomeness aside, he found that when he spoke into the ear, the hay traced the sound waves on a piece of smoked glass.

Bell would draw on these early observations when he set about inventing his telephone. Initially he designed a series of reeds arranged over a long magnet. As each reed responded to the voice, it would vibrate, creating the undulating current. The design was similar to piano strings, each of which corresponds to a specific note or "pitch." In June 1875, Bell asked his assistant, Thomas Watson, to pluck a steel receiver reed with his finger to make sure it wasn't stuck, as part of a routine test while setting up an experiment. When Watson vibrated the reed, the receiver in Bell's room also vibrated, even though the current wasn't turned on. He then realized that a single reed could convey all the elements of human speech.

Bell promptly constructed a prototype telephone in which the reed was attached to a membrane with a speaking cavity positioned above it, but this did not produce intelligible speech, apart from a low mumbling. Still, it was enough to convince him that he was on the right track, and he submitted a patent for the device on February 14, 1876—barely edging out Elisha Gray, who submitted his own design for a speaking telegraph a few hours later. One month later, Bell revised his design. This new version included a speaking tube and membrane using a cork to attach a needle as a vibrating contact. One of his reed receivers was placed in another room, and Bell then spoke the now famous words: "Watson, come here; I want to see you."

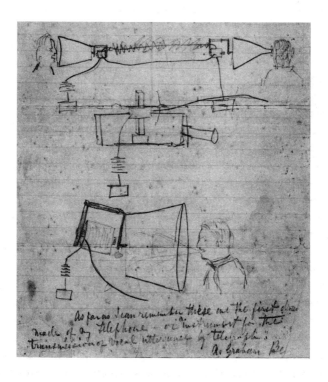

Bell's design sketch for a telephone

This was the first "documented" transmission of human speech over a telephone wire. The distinction is important because an Italian immigrant named Antonio Meucci had invented a telephonic device in 1849, long before Bell and Gray. Meucci studied design and mechanical engineering before becoming a scenic designer and stage technician. He was fascinated by science and conducted many experiments, even developing a method of electric shocks to treat illness. One day, while preparing to administer a treatment to a friend, Meucci heard his friend's voice over the piece of copper wire running between them, even though the friend was in the next room. He spent the next 10 years trying to build a practical device, emigrating to New York City in 1850 to pursue that end. In 1855, his wife became partially paralyzed; Meucci set up a telephonic intercom at his home in Staten Island, joining several rooms with his workshop, which was in a nearby building.

Meucci was beset by poverty and illness for most of life. When he was severely burned in a steamship explosion, his wife sold his telephone prototype and several other models to a secondhand dealer for $6. Meucci sought to recover the items, but was told they had been sold to "an unknown young man." He reconstructed his telephone prototype, and although he couldn't afford the $250 patent fee, he did file a notice of intent on December 28, 1871. Meucci even went so far as to deliver one of his models, with accompanying technical details, to the vice president of one of the Western Union Telegraph Company's fledgling affiliates. He wanted to demonstrate the feasibility of his invention—which he called a "talking telegraph"—over the wires of the company's telegraph network. But the vice president was maddeningly noncommittal every time Meucci contacted him subsequently over the next two years. Finally, fed up with being given the runaround, Meucci demanded the return of his prototype in 1874, only to be told it had been "lost."

When Bell filed his patent for a telephone, Meucci hired a lawyer, and learned that all prior documentation he had filed had also been mysteriously lost. Historians now suspect there may have been illegal relationships between employees of the U.S. Patent Office and those at Bell's company. In a high-profile court case in 1886, Meucci explained every detail of his invention so clearly that there was little doubt he was telling the truth. But he still lost the case against the far wealthier and influential Bell. Meucci died 10 years later. His claim wasn't recognized until June 15, 2002, when Congress officially credited Meucci with the invention of the telephone.

The controversy has somewhat tarnished Bell's image, but there is no doubt that he was an ingenious inventor. He unveiled his telephone at the Centennial Exhibition in Philadelphia in 1876 to an awestruck crowd, prompting Emperor Don Pedro of Brazil to exclaim, "My God, it talks!" And in June 1880, Bell transmitted the first wireless telephone message on his newly invented photophone—13 years before the Serbian inventor Nikola Tesla would give the first public demonstration of wireless technology using radio waves. Unlike the telephone, which relies on electricity, the photophone transmitted sound on a beam of light. Bell's voice was projected through the instrument to a mirror, causing similar vibrations in the mirror. When he directed sunlight into the mirror, it captured and projected the mirror's vibrations as reflections, which were then transformed back into sound at the receiving end of the projection.

Bell's device did not find immediate application, but today scientists recognize it as the progenitor of modern fiber-optic telecommunications. Every day, fiber-optic systems send millions of telephonic signals as pulses of light all over the globe—and perhaps the occasional lapsed demon, beaming himself away from the hordes of Hades.

14

Thrill Seekers

1884
First U.S. roller coaster opens at Coney Island

One bright spring morning in 1999, a crowd gathered at Busch Gardens in Williamsburg, Virginia, for the opening of the park's new roller coaster, Apollo's Chariot. On hand as a special celebrity guest was the male model Fabio—he of the flowing blond locks, chiseled jaw, and impeccably sculptured torso. Best known for posing in strategically ripped shirts on the covers of mass-market romance novels, and for hawking butter substitutes on television, Fabio had the Apollonian good looks that made him an obvious choice to usher in the new ride.

But halfway through the initial 210-foot drop, disaster struck: a wild goose flew into the coaster's path and smashed into Fabio's face. The impact gashed Fabio's nose and killed the goose, whose broken body was later fished out of a nearby river. Fabio ended the ride with his face covered in blood (whether his own or that of the goose, no one could say)—a gruesome testament to the impressive forces a simple thrill ride can generate.

Affectionately dubbed the "great American scream machines," roller coasters derive from the giant ice slides that first appeared in the seventeenth century in Russia, near Saint Petersburg. The slides were built out of lumber covered with a sheet of ice several inches thick, and featured drops of 50 to 80 feet. Riders would walk up a ladder or set of stairs to the top of the slide, climb into wooden sleds, and shoot down the slope, crashing into sand piles at the base of the slope to stop the ride. The slides were a favorite diversion of the Russian upper class.

Catherine the Great was such a fan that she had a few built on her estate.

Roller coasters have come a long way since those rudimentary ice slides, but both rely on the same underlying principles of physics: inertia, gravity, and acceleration. A roller coaster is always shifting between potential (stored) and kinetic (released) energy. Compressing a spring stores potential energy in the coils, which is released as kinetic energy when the spring pops back out to its original shape. Roller coasters build up a large reservoir of potential energy while being towed up the initial "lift hill." Initially they were towed manually; modern roller coasters are pulled up lift hills by chain-and-pulley systems, powered by small electric motors. After that initial tow, physics takes over.

The key factor at play is gravity, which applies a constant downward force on the coaster's cars; the tracks channel this force by controlling the way the cars fall. The higher the coaster train rises, the greater the distance gravity must pull it back down, and the greater the resulting speeds. As the train starts down the first hill, the potential energy is converted into kinetic energy and the train speeds up, building up enough kinetic energy by the time it reaches the bottom to overcome gravity's pull and move up the next hill—and so on for the rest of the ride. But the laws of thermodynamics dictate that no machine can operate indefinitely without getting an extra influx of energy from somewhere. Roller coasters are no exception. That's why the hills in most modern coasters tend to decrease in height as the ride progresses; the total energy reservoir built up on the lift hill is gradually lost to friction between the train and the track—entropy in action.

Most historians credit the French with building the first wheeled coasters. Although the French loved the concept of the Russian ice slides, the warmer climate in France would melt the ice for all but a few months of the year. So they started building waxed slides instead, and to help the sleds move down the slides, they added wheels. Later, they substituted a wooden track for the waxed slide. By 1817 there were two bona fide roller coasters in France, both of which featured cars locked to a track through wheel axles fitted into a carved groove.

In 1846, the world's first looping coaster—in which riders are turned upside down—was unveiled at Frascati Gardens in Paris. The ride featured a 43-foot lift hill and a 13-foot-wide loop. The speeds were slow compared with those of modern rides, but the basic physics remained the same. When a coaster loops the loop, centripetal force comes into play. It's the same underlying principle as in the merry-go-round: the forces generated by the spinning platform push riders out and away from the center of the platform, but those forces are countered by a restraining bar accelerating riders back toward the center.

A roller coaster loop-the-loop is, in essence, a merry-go-round turned on its side—although most modern loops are teardrop-shaped (not perfect circles), to better balance the various forces acting on the body. At the bottom, the acceleration force pushes riders down in the same direction as gravity, so they feel especially heavy. As the ride moves up the loop, gravity pulls the riders into their seats while acceleration pushes them into the floor. At the top, when the riders are completely upside down, gravity pulls them out of their seats toward the ground, while the stronger acceleration force pushes them into their seats toward the sky; the two forces effectively cancel each other out. As a result, the riders experience a momentary feeling of weightlessness or "free fall," suspended in midair for a split second.

An early attempt to bring roller coasters to the United States failed because of an accident during the trial run. Then along came an American inventor, La Marcus Thompson, who would ultimately revolutionize the amusement industry in the United States. Thompson was born in 1848 in Jersey, Ohio, and was a natural at mechanics, designing and building a butter churn for his mother and an oxcart for his father when he was only 12. After finishing college, he worked in the wagon and carriage business before making his fortune as a manufacturer of women's hosiery. After selling his stake in the hosiery business, he turned back to his first love: inventing.

When he was in the manufacturing business, Thompson had ridden on the Mauch Chuk Switchback Railway in Pennsylvania, a former mine track that had become a popular tourist attraction. The track was originally built as an easy way to send coal down the mountain to the railway 18 miles away. But then the railway built a new tunnel, bringing the freight trains much closer to the coal mine. So the switchback railway was reconfigured as a "scenic tour." For $1, tourists got a leisurely ride up to the top of the mountain, followed by a wild bumpy ride back down.

Thompson decided to build his own Gravity Pleasure Switchback Railway at Coney Island in New York, the previous location of a track for horse races. Completed in 1884, it was the first roller coaster to be built in the United States. It incorporated undulating hills and a flat steel track nailed onto several layers of wooden plank, connected to two towers. The maximum speed was a poky 6 miles per hour, and the cars had to be manually towed to the top of the hills at the start of both tracks. Still, the ride was an instant success with the public, who had little to compare it with, and Coney Island mushroomed into a full-scale amusement park. No one knows when the rides became known as "roller coasters"; they were initially called switchback railways or centrifugal railways. Thompson received a U.S. patent for his "rolling coaster" ride in January 1885. His wasn't the first. Three other similar patents were issued in 1884, and some historians believe that Thompson actually based many features of his design on a patent issued in January 1878 to Richard Knudson for "improvements in inclined railways."

Within four years, Thompson had built approximately 50 more such coasters across the nation and in Europe and then designed his most famous attraction, the Scenic Railway. It opened in 1887 in Atlantic City and featured artificial scenery illuminated by lights triggered by the approaching cars—a precursor of the elaborate rides at Disneyland and other modern theme parks and amusement parks.

Improvements and innovations came fast on Thompson's heels. Later in 1884, Charles Alcoke designed a coaster with a continuous track, so that the ride ended where it began, and in 1885 Phillip Hinckle was the first to use a mechanical hoist to raise the cars to the top of the lift hill. That same year saw the debut of a coaster called the Flip-Flap, which rolled cars through a loop-the-loop 25 feet in diameter, but it closed in 1903 because the passengers frequently suffered neck and back injuries. The ride was slow but far from smooth, and the sharp, jerking motions as passengers were flipped upside down often resulted in severe whiplash.

By the end of the nineteenth century, all the basic elements of the modern roller coaster were in place, although the coasters were still frustratingly slow. The early 1900s brought numerous innovations in roller coaster design that increased the speeds by better manipulating the various forces: higher lift hills, sharper turns, and sharp spiraling drops. The 1920s were the golden age of roller coaster design, with more than 1500 rides opening in North America alone. But the Great Depression put an end to such frivolity, and many coasters were torn down. From 1930 to 1972, only 120 roller coasters were built in the United States, while more than 1500 were destroyed.

The technology languished until 1955, when the opening of Disneyland in southern California marked the introduction of tubular steel tracks. The earlier coasters were wooden, similar to traditional railroad tracks. They were too inflexible to enable designers to construct complex twists and turns, and the cars tended to rattle as they rolled over the joints connecting the track. The new tracks were made of long steel tubes supported by a superstructure made of slightly larger tubes, and the pieces were welded together. Not only did this make for a smoother ride, but tubular steel coasters allowed more looping, higher and steeper hills, greater drops and rolls, and faster speeds.

This led to a second roller coaster boom in the 1970s and early 1980s, which revitalized the industry with the introduction of all sorts of innovative designs, like the corkscrew track. Today, catapult launch systems, suspended train designs (in which the riders' feet dangle, and the "cars" swing from side to side), elaborate ride themes, and other twists have come fast and furiously, making roller coasters more popular than ever. There are coasters in which you lie flat against the train car and have the sensation of flying, and coasters that shoot you down long stretches of spiraled track. In 1997, a coaster opened at Six Flags Magic Mountain in California that was 415 feet tall and could reach record-breaking speeds of 100 miles per hour.

Thanks to the continual variation in forces, roller coasters put the human body through a number of exhilarating physical sensations in a matter of seconds. These are the so-called "g forces," which indicate how much force the rider is actually feeling. The riders' inertia is separate from that of the car, so when the car speeds up, they feel pressed back against the seat because the car is pushing them forward, accelerating their motion. When the car slows down, a rider's body tends to continue forward at the same speed in the same direction, but the harness or restraining bar decelerates it—that is, slows it down.

A "g" is a unit for measuring acceleration in terms of gravity. It also determines how much we weigh, as opposed to our mass (how many atoms make up our body). Weight is determined by multiplying an object's mass by the force of Earth's gravity. The g forces arise because a roller coaster is accelerating: forward and backward, up and down, and side to side. This produces corresponding variations in the strength of gravity's pull. For example, 1 g is the force of Earth's gravity: what the rider feels when the car is stationary or moving at a constant speed. Acceleration causes a corresponding increase in weight, so that at 4 g, for example, a rider will experience a force equal to four times his weight.

At high speeds, the g forces can be considerable. Fabio endured a lot of ridicule in the media after his encounter with the kamikaze goose; people were amused that the 6-foot 3-inch, 220-pound hunk fared so poorly against a 22-pound waterfowl. But assuming the collision lasted a hundredth of a second, and the coaster was traveling at a speed of about 70 miles per hour, Fabio would have absorbed an impact equivalent to a hard tackle by the football player Mean Joe Green, delivered with a force equivalent to a solid punch from the heavyweight champ Mike Tyson. Yet not one reporter said, "That Fabio, he can really take a punch!"

So there is a dark side to all this merry mayhem: accidents and injuries do happen, and coaster-related (human) deaths number between two and four per year. Compared with the hundreds of millions of visitors who crowd amusement parks every year, this might seem insignificant; fatalities occur for about one in 450 million riders. But the statistical anomaly is small comfort to the victims and their families. The federal Consumer Products Safety Commission reported in 1999 that there had been an 87 per cent increase in injuries on amusement park rides from 1994 to 1998; it attributed the increase in part to the steadily increasing acceleration forces generated by the rides. The newest coasters can reach top speeds of 100 miles per hour with g ratings as high as 6.5. For comparison, astronauts typically experience 4 g while traveling up to 17,440 miles per hour on liftoff, and NASCAR drivers have reported feeling dizzy after experiencing

5 g. Coaster designers counter this by pointing out that astronauts and NASCAR racers experience sustained g forces; riders on a roller coaster are typically exposed to high g forces for only 1 second or less.

Mechanical failure is sometimes the culprit, but some of the most spectacular accidents occur because riders ignore basic safety precautions. Thrill seekers have been known to remove the safety harness; this act can chuck riders out of the car and send them flying through the air at high speeds. Standing up during the ride has also caused numerous riders to fly out of cars, or to strike their heads on low beams, usually suffering fatal injuries. In 1996, at Six Flags Great America, a man wandered into a restricted track area to retrieve his wife's hat, which had blown off in the high winds. A rider on the Top Gun suspension coaster kicked him in the head, killing him instantly. The rider suffered a broken leg. Six years later, a rider on the Batman suspension coaster at Six Flags Over Georgia leaned out of the car and nearly lost his head when a rider in a train traveling in the other direction on an adjacent track kicked it. The man who leaned out of his car was killed immediately by the impact. And in a bizarre incident in May 2003, an 11-year-old girl choked to death on her chewing gum while riding a coaster at Six Flags Great America.

More insidious are the injuries many riders may not even be aware they've suffered; some doctors believe that the sharp jerks and jostles of high-speed rides could have the same brain-battering effects as professional football. The strong g forces generated by high-speed coasters can cause headaches, nausea, and dizziness—possibly harmless, but also symptoms of mild concussion—simply because the body doesn't have sufficient time to adapt to the rapidly changing environment. The effect can be similar to what happens to the brain during a car accident, or when a person is violently shaken. As the head whips sharply back and forth, the brain can pull away from one side of the skull and smash into the other side with sufficient force to rupture tiny blood vessels. The trickling blood accumulates in the small space between the brain and the skull, and the resulting pressure can lead to permanent brain damage or death if left untreated. In the summer of 2001 alone, three women suffered fatal brain injuries on roller coasters in California, although two of those victims had a preexisting aneurysm—a weak spot on a blood vessel in the brain—which ruptured during the ride.

Some of these physical effects are admittedly speculative, and none are likely to dissuade any diehard coaster fans, for whom the potential dangers are part of the thrill. But Fabio's encounter prompted him to call for increased safety measures on roller coasters. And he isn't the only naysayer. Concerned doctors, legislators, and consumer groups continue to lobby for tighter regulation of the amusement park industry, in hopes of preventing similar fowl incidents, thus ensuring the future well-being of children and wild geese everywhere.

15

Current
Affairs

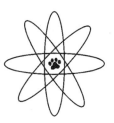

On August 14, 2003, unsuspecting residents across the northeast were going about their daily activities when a major power outage struck simultaneously across dozens of cities: New York, Cleveland, Detroit, several towns in New Jersey, and even Toronto, Canada, were among those affected. In 3 minutes, 21 power plants shut down, leaving some 50 million people without power. The outage stopped trains, trapped people in elevators, and disrupted the normal flow of traffic. In Michigan even the water supply was affected, since water in that state is distributed through electric pumps. In New York City, it took 2.5 hours to evacuate passengers from stalled subway trains, and many commuters were stranded, although some residents of New Jersey were ferried home across the Hudson River. The lack of air conditioning on that typically hot, humid summer evening proved to be the least of anyone's worries.

Speculation abounded as to the cause of the outage, but it was due to neither a terrorist act nor the so-called "Blaster worm," a computer virus then spreading rampantly over the Internet. Early indications pointed to a downed 345,000-volt power line east of Cleveland as the source of the outage. The 21 plants went offline for the very simple reason that when the power distribution grid is down, there is no place for the energy output to go. Electricity, unlike fuels such as coal or natural gas, is difficult to store, so electric power is generated only as it is used.

The nation's power grid boasts more than 6000 interconnected power generation stations. Power is zapped around the country via 500,000 miles of bulk transmission lines carrying high-voltage charges of electricity. From these lines, power is sent to regional and neighborhood substations, where the electricity is then stepped down from high voltage to a current suitable for use in homes and offices. The system has its advantages: distant stations can provide electricity to cities and towns that may have lost power. But unusually high or unbalanced demands for power—especially those that develop suddenly—can upset the smooth flow of electricity. This can cause a blackout in one section of a grid, or ripple through the entire grid, shutting down one section after another. That's what happened in the blackout of August 2003: large sections of the grid were shut down, making it difficult to restore power from neighboring stations.

There are two kinds of current electricity: alternating current (AC) and direct current (DC). In direct current a steady stream of electrons flows continuously in only one direction: for example, from the negative to the positive terminal of a battery. Alternating current changes direction 50 or 60 times per second, oscillating up and down. Almost all the electricity used in homes and businesses is alternating current. That's because it's easier to send AC over long distances without losing too much to leakage. Leakage is the result of friction as electricity travels along a wire over a distance; some voltage loss inevitably occurs. AC can be converted much more easily to higher voltages, which are better able to overcome line resistance. If DC current were used for power transmission, we would need new generating stations roughly every 20 feet to beef up the voltage for the next leg of its journey.

We owe much of the development of AC power to the nineteenth-century inventor Nikola Tesla, who was actually born during an electric storm in Lika, Croatia. Tesla was the son of a Serbian orthodox priest, who hoped his son would follow in his footsteps. But Tesla had a gift for science and invention—the latter most likely inherited from his mother, who was skilled at creating her own useful household gadgets. As a student in Austria, Tesla became intrigued by electricity, a fascination that would last his entire life. At the time, most motors ran on DC power; and attempts to create an AC version had failed. In 1882, while working as an electrical engineer with a Hungarian telephone company, Tesla conceived of a rotating magnetic field produced by two or more alternating currents out of step with each other. Margaret Cheney, in her definitive biography of Tesla, *Man Out of Time*, described it as "a sort of magnetic whirlwind" that could be used to drive an electric motor. This was a vital missing piece for a fully operational AC power generating and transmission system, since it is what creates a steady flow of electrons along the conducting wires. Within a few months Tesla had sketched out plans for all the components of just such a system.

Tesla emigrated to the United States in 1884 and immediately found work in Thomas Edison's lab in Menlo Park, New Jersey. Edison admired Tesla's drive and extraordinary work ethic—the Serb often worked through the night—but the two men soon found themselves at odds. That was partly due to vast differences in personality. Tesla was cultured, educated, soft-spoken, and fluent in several languages, with a solid understanding of the electromagnetic theory underlying his inventions. Edison was brash, a bit uncouth, and openly contemptuous of scientific theory; if it didn't make money, it was of no value to him. Edison dubbed Tesla "the poet of science," and considered his ideas "magnificent but utterly impractical."

The real cause of the animosity was Edison's brusque dismissal of Tesla's idea for an AC system of power generation—largely because Edison had already invested heavily in promoting his own DC system. This was the height of the so-called "current wars," when hundreds of central power stations were cropping up across America, each using different combinations of circuits and equipment. Edison's Pearl Street generating station in Manhattan supplied DC power to a few hundred mansions of wealthy New Yorkers, as well as a smattering of mills, factories, and theaters in the city. His archrival, George Westinghouse, espoused AC and had more than 30 AC plants in operation by 1887. Whoever emerged victorious stood to gain a veritable monopoly on a highly lucrative market. When Tesla quit his job with Edison—in a huff, after being denied a promised bonus—Westinghouse was waiting in the wings. He offered Tesla a generous salary of $2000 a month, plus substantial royalties on his various AC patents.

Edison was notoriously ruthless when it came to quashing rivals. Facing resistance from gas companies when he first introduced DC power, he fought back by publishing regular bulletins on the dangers of gas-main explosions so as to induce fear in the public mind. The strategy worked, and he adopted a similar ploy to combat the Tesla-Westinghouse AC system, launching an intensive smear campaign. The Edison propaganda machine pumped out hundreds of flyers and pamphlets on the dangers of alternating current, and if genuine accidents could not be found—well, they could always be staged. Soon families living near Edison's lab in New Jersey noticed that their pets were disappearing. Edison was paying young boys to bring in dogs and cats, which he strapped onto a sheet of metal connected by wires to an AC generator. The animals were then electrocuted with 1000 volts of alternating current in front of any newspaper reporters able and willing to stomach the spectacle. Edison called this being "Westinghoused."

Even more deviously, he recruited a former lab assistant, Harold Brown, to obtain licenses for three of Tesla's AC patents under false pretenses. He then persuaded officials at Sing-Sing prison to carry out future executions not by hanging, but by electrocution using alternating current. On August 6, 1890, a convict named William Kemmler was strapped into the very first electric chair, and the switch was thrown. Unfortunately for Kemmler, Edison's engineers had grossly underestimated the amount of voltage needed to kill a man; all their experiments had been conducted on much smaller animals. The grisly procedure had to be repeated before the condemned man finally died. A reporter who witnessed the execution deemed the experience "far worse than hanging."

Tesla was also known to electrocute small animals as part of his demonstrations, but his aim was to show that AC could be safely controlled. He would stand upon the same platform on which the animal had met its demise, and proceed to conduct as much as 2 million volts of current through his own body, until his lanky frame was ringed with a halo of electricity. One observer described him as a "human electric live wire." When asked why he hadn't been electrocuted, he explained that alternating currents of high voltages would flow on the outer surface of the skin without causing injury if the frequencies were high. It was low-frequency currents flowing beneath the skin that were lethal, whether AC or DC.

Tesla clearly loved electricity, and he understood its behavior on an almost intuitive level. And it seemed to love him in return. He quickly usurped Edison's place as the "wizard of electricity." His public lectures were hugely popular. He stood nearly 7 feet tall on the lecture platform, since he always wore thick cork soles on his shoes for protection during demonstrations. Before awestruck crowds, Tesla shot sparks from his fingertips, made lightbulbs glow, and even melted metals by running current through his body. He was able to snap his fingers and create a ball of red flame, holding it calmly in his hand without being burned. Some of his demonstrations have never been duplicated. His fearlessness did a great deal to counter Edison's smear campaign and allay the public's fears about AC.

In 1893, the Columbian Exposition in Chicago—commemorating the four-hundredth anniversary of America's discovery—chose AC to light up its White City, the centerpiece of its World of Tomorrow exhibit. That successful demonstration was sufficient to win Westinghouse the contract to build the first two AC generators at Niagara Falls. So Westinghouse won the current wars. By the late 1920s, more than $50 billion had been invested in Tesla-Westinghouse AC power transmission systems. This should have made Tesla one of the wealthiest men in the world, but his idealism proved to be his undoing. As AC power spread across the country, the patent royalties promised to him by Westinghouse mounted so quickly that paying them in full would have driven the company into bankruptcy. Tesla would be rich—but at the expense of his own invention. When told of the predicament, Tesla barely hesitated: he tore up the royalty contract and relinquished his claim to a fortune. His reasoning was that Westinghouse had believed in him when others had nothing but ridicule and contempt for his ideas. It was an astonishingly selfless act, ensuring the success of the company and the continued dominance of AC power. Tesla, a modern-day Prometheus who wanted only to give electricity to the world, ended his days impoverished.

As efficient as the AC system is, blackouts still occur periodically. Before August 2003, the great northeastern blackout of November 1965 was the largest in U.S. history, depriving at least 25 million people of electricity in New York, New England, and portions of Pennsylvania and New Jersey. Another major blackout occurred in August 1996, affecting some 4 million customers in nine western states and parts of Mexico. Tesla himself caused a blackout in 1899. He was conducting an experiment at his laboratory in Colorado Springs that required millions of volts—currents so powerful that snakes of flame and lightning bolts shot 135 feet in the air. Tesla watched the spectacle from his doorway, transfixed with joy, when suddenly everything stopped. The power was dead. Tesla called Colorado Springs Electric Company, only to be told that his experiment had knocked their generator offline and set it on fire. The entire town was in darkness. The company refused to restore his power until he had repaired its generator, free of charge.

When it comes to modern blackouts, however, the real culprit is deregulation. Well before the outage of August 2003, analysts had been warning of just such an occurrence. In 1998, a former utility executive named John Casazza predicted that the risk of blackouts would increase if the government went ahead with deregulation, and his warnings were echoed by many other energy experts. Legislators didn't heed the warnings.

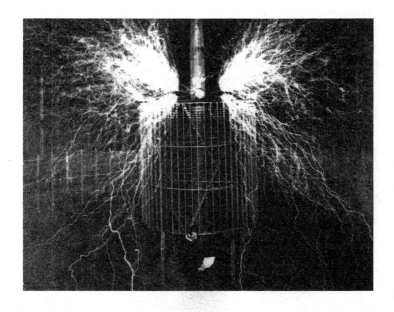

**Tesla in his laboratory
in Colorado**

Before deregulation, a single company controlled power generation, transmission, and distribution in a given geographical area, with enough capacity to meet its customers' needs. Long-distance energy shipments were reserved for unexpected outages. This made the system more reliable, because the greater the distance electricity has to travel, the more complex the interconnections, and the more vulnerable the system becomes to an outage. Just as lightning will follow the path of least resistance rather than a straight line, power flows from point A to point B through any number of available interconnects, depending on which has the most available capacity. If certain parts of the grid are running near capacity, the situation can create the electrical equivalent of a traffic jam during rush hour. Even a small bit of additional power can overload the wires, tripping circuit breakers and causing a chain-reaction failure. "Traffic" comes to a standstill.

With deregulation, power generation was separated from transmission and distribution, and treated like a traded commodity instead of an essential service. Generating companies sell their power for the highest price they can get, and utilities buy it at the lowest possible price, regardless of the seller's geographic location. While this might seem perfectly reasonable in a free-market economy, it ignores the fundamental physics of how electricity behaves. Making matters worse, utilities began withholding vital data on energy flows, claiming that these flows were privileged information. As a result, the reliability of the system can't be accurately assessed. It's like having individual musicians in an orchestra play their own tunes while hiding their sheet music from the other musicians. The result is a system that is simultaneously less coordinated and more difficult to control.

It didn't help that new energy companies such as Enron, Reliant, and Dynergy essentially "gamed" the power grid system, running transmission lines near full capacity to squeeze as much money as possible out of trading. Some companies tried to stack the deck: federal investigators found that Enron and Reliant had deliberately blocked competitors' access to the grid to jack up prices by creating artificial shortages in California. The state suffered widespread blackouts, and Californians saw their electricity rates double and even triple as demand quickly outstripped supply. Since the utilities had sold off so much of their generating capacity, they were at the mercy of these price-gouging privately held companies.

Deregulation still has its proponents, who cling to their conviction that the practice will eventually lead to more efficient systems and lower average costs to consumers, despite mounting evidence to the contrary. They argue that California is an extreme case. But that doesn't explain why two power companies in California were immune to the crisis. Both were owned by the public, and their rates remained stable.

Other areas have fared better than California under deregulation. But deregulation still makes the power grid system subject to the whims of market supply and demand. In densely populated states like New York and Massachusetts, power rates are among the highest in the country. In 2002, residents of New York paid an average of 13.58 cents per kilowatt-hour. In contrast, residents of Kentucky paid less than half that amount: 5.65 cents per kilowatt-hour. True, the cost of living in general is significantly less in Kentucky than in New York, but the whole point of deregulation, ostensibly, was to bring savings to all consumers.

In that regard, the practice has been an abject failure. Mark Cooper, director of research at the Consumer Federation of America, has described the effect of deregulation as ranging "from nothing to unmitigated disaster." Despite lawmakers' promises, deregulation has produced no savings for American consumers—just substantial rate increases, more frequent blackouts, and huge profits for unscrupulous private generating companies. The laws of physics notwithstanding, apparently corporate greed still reigns supreme. Where's an idealistic inventor when you need one?

16

Shadow Casters

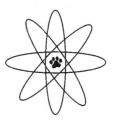

February 2, 1893
Edison films a sneeze

Toward the end of the sixteenth century, an Italian scientist named Giambattista della Porta nearly caused a riot when he seated some guests inside a darkened room and positioned actors outside a window to perform a short vignette. The sunlight shining through the window cast upside-down moving images of the actors on the viewing wall. It was meant as light entertainment. Instead, his guests panicked at the sight of the spectral figures and fled, and Giambattista was brought to court on charges of sorcery (which were later dismissed). He was hardly the first to cast spectral images—illusionists had been doing it for centuries—but he was still a little too far ahead of his time: it would be another 300 years before the underlying optical effect behind the trick gave rise to modern cinematography.

Today, millions of people flock to movie theaters every weekend to view the latest Hollywood blockbusters, but few stop to consider the technological roots of this multibillion-dollar entertainment industry. All forms of photography date back to an ancient optical effect called the camera obscura (described in greater detail in Chapter 7), in which inverted images of external objects are formed on a white surface within a darkened chamber or box. First used to observe solar eclipses and later as an optical sketching aid for artists, the camera obscura also provided a unique form of entertainment. Around 121 BC, in China, a magician named Shao Ong performed a shadow play in which he claimed that he made the spirit of a dead concubine appear to Emperor Wu. And Ptolemy discussed "stereoscopic projection" in his treatise *Almagest* (c. AD 140).

In 1290, Arnau de Villanova, an alchemist and a practising physician who was also a magician in his leisure time, used a camera obscura to present "moving shows" or "cinema." Like Giambattista, he would place his audience in a darkened room and have the actors perform just outside, and the images of the performance—usually enactments of wars or hunts—would be cast, upside down, on the inside wall. His audiences were far less reactionary than Giambattista's, however, and he avoided charges of sorcery. By the nineteenth century, such performances had become so popular that traveling magic shows regularly used them to entertain audiences.

One of the most famous variations on spectroscopic projection is known as "Pepper's Ghost." In 1862, an English showman, "Professor" Henry Pepper, collaborated with a retired civil engineer named Henry Dircks on a concept for presenting an actual "ghost" on a theater stage.

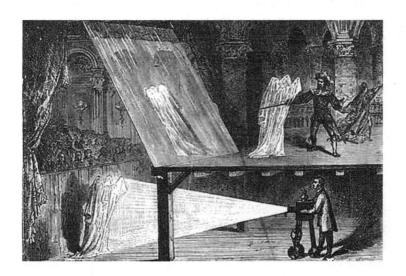

Pepper's ghost

They used a large piece of plate glass, tilted at a 45-degree angle. When offstage objects were illuminated with bright light, their images would be superimposed onto the stage. An actor dressed as a ghost was positioned offstage and was illuminated by a lantern. His transparent image, formed on the angled sheet of glass, would appear to the audience to "interact" with live performers. Later versions included a lens to focus and improve the image.

It was a neat little trick that no doubt added verisimilitude to nineteenth-century productions of *Hamlet* and *Macbeth*, although there were some inherent difficulties with staging. In a pre-microphone age that relied on natural means of amplification, the glass would partially block sound waves, making the actors difficult to hear. But the effect was hugely successful with traveling carnivals. Pepper himself took his stage illusion on tour around the world, even to Australia. There, his act was inspired by a local ghost story about a paroled convict, Fredrick Fisher, who had disappeared in 1826. Fisher had returned home to find his farm claimed by a friend, George Worrell. Then Fisher mysteriously disappeared, and Worrell began wearing Fisher's clothing and claiming to possess a receipt for purchasing the farm. Fisher's body was finally found when a man named James Farley, walking along the property one night, encountered a ghostly figure that pointed toward the creek—Fisher's final resting place.

The concept of the camera obscura is also the basis for the pin-hole camera and the emergence of photography, a development that was eerily prefigured in a mid-eighteenth-century science-fiction novel by Charles François Tiphaigne de la Roche called *Giphantie*. He envisioned an imaginary world where it was possible to capture an image from nature on a canvas coated with a sticky substance, which, after being dried in the dark, would preserve the image. It would take more than a century for chemistry to catch up with de la Roche's imagination. Scientists already knew that silver chloride and silver nitrate both turned dark when ex posed to light, and the first silhouette images were captured by Thomas Wedgwood at the start of the nineteenth century. But it still hadn't occurred to anyone that this photochemical effect could be used to make images permanent.

Photography essentially freezes a moment in time. As light bounces off the objects in the camera lens field of view, it catalyzes a reaction in the chemicals coating the film inside the camera—small grains of silver halide crystals. These crystals are naturally sensitive to light. By opening a camera's shutter for a split second, you expose the crystals to light and transfer energy from the photons to the silver halide crystals. This induces the chemical reaction, forming a latent image of the visible light reflected off the objects in the viewfinder. If too much light is let in, too many grains will react and the picture will appear washed-out. Too little light has the opposite effect: not enough grains react, and the picture is too dark, as anyone who has ever taken an indoor photo without a flashbulb can attest. Changing the size of the aperture or lens opening controls the amount of light. In modern cameras, this is the job of the diaphragm, which works the same way as the pupil in the eye. Chemicals are used in the developing process, which react in turn with the light-sensitive grains, darkening those exposed to light to produce a negative, which is then converted into a positive image in the printing process.

Vera Lutter, an artist in New York City, has been creating "photographs" for years using her own camera obscura. She built a wooden shack on the roof of the landmark Pepsi-Cola factory in Queens in 1998 and spent the next six years making prints of the giant red neon Pepsi sign, which was first erected in 1936. Sunlight would stream through a small opening on one side of the shack and project an upside-down, reversed image of the Pepsi sign on the opposite wall. Each morning, Lutter draped three big sheets of photosensitive paper on that wall. The light burned the image into the paper over the course of three hours—the minimum exposure time needed to get a usable image—and she then developed and assembled the images into single 15- by 8-foot photographic prints at her studio in Manhattan. The Whitney Museum is among the venues that have exhibited her prints. The Pepsi-Cola sign has since been relocated; its dismantling has been carefully documented by Lutter for posterity.

The Frenchman Joseph-Nicéphore Niepce took the first still photograph in July 1827, using a material that hardened when exposed to light to capture the image, but it required a full eight hours of exposure, and the image was temporary. Six years later, Louis Daguerre discovered how to develop photographic plates, reducing exposure time to half an hour, and managed to make the images permanent by immersing the plates in a salt solution. These "daguerrotypes" were the earliest form of still photography and became hugely popular. Renowned figures as diverse as President Abraham Lincoln and the poet Emily Dickinson had their images captured for posterity in daguerrotypes, and the process enabled the first photojournalists to document the horrors of the American Civil War. But daguerrotypes were expensive, and no extra prints could be made. The only way to produce copies was to use two separate cameras side by side. So an Englishman named William Henry Fox Talbot invented a rival technology, the calotype, which produced paper negatives of poorer quality than the daguerrotypes but had the capability to produce an unlimited number of positive prints. Modern photography is based on the same principle.

Photography studios began springing up throughout Europe in the 1840s. By the mid-1860s, Regent Street in London had 42 photography studios; in America, there were 77 in New York alone by 1850. It became standard practice to include photographs on calling cards (the calling card was required by upper-class etiquette of the time). And inventors continued to improve the process. Several experimented with glass as a basis for negatives, but the silver solution wouldn't stick to the shiny surface. By 1848, Abel Niepce de Saint-Victor came up with the idea of coating a glass plate with egg white mixed with potassium iodide, and then washing it with an acid solution of silver nitrate. The result was fine detail and vastly higher quality, but once again the procedure required prolonged exposure. Three years later Frederick Scott Archer introduced the collodion process, which reduced exposure time to a few seconds; but this was still a "wet" process, requiring that all the equipment be on-site at the time the picture was taken. Finally, in 1871, Richard Maddox found a way of using gelatin instead of glass as the basis for the photographic plate; he developed a dry plate process that could produce photographs much more quickly.

It was only a matter of time before inventors began speculating about ways to make the pictures move. People were already fashioning crude hand-drawn motion pictures, similar to early animated cartoons, or to the flip books we played with as children. The British photographer Eadweard Muybridge took that concept one step farther and helped pioneer a process in which a series of pictures would be taken of a subject in motion and then shown in sequence. Muybridge was interested in studying movement, and was asked to settle a bet for the governor of California, who insisted that when a horse gallops, at a particular point all four feet are off the ground simultaneously. Aside from one brief distraction—Muybridge was tried and acquitted for killing his wife's lover, in the O. J. Simpson trial of the nineteenth century—he devoted himself to the task, and succeeded in photographing a horse galloping, using 24 cameras connected to trip wires on the course. The images proved the governor right, and took public attention away from Muybridge's scandalous trial.

Muybridge continued to conduct comprehensive photographic studies of men and women in motion, and in 1878, *Scientific American* published some of these photographs and suggested that readers cut out the pictures and place them in an optical toy called a "zoetrope" to re-create the illusion of movement. The zoetrope was essentially a small drum with an open top, supported by a central axis around which it spun. Sequences of hand-drawn images on strips of paper were placed around the inner wall of the drum, and slots were cut at equal distances along the sides, all around the drum's outer surface—just above where the pictures were positioned. This enabled people to view the images in motion as the drum was spun. The faster the rate of spin, the more fluid the progression of images.

Intrigued by the concept, Muybridge invented his own version, the zoopraxiscope, which projected images from rotating glass disks in rapid succession. The images were painted directly onto the glass. This was essentially the first film projector, and it quickly caught the attention of American inventor Thomas Edison—the "wizard of Menlo Park"—who had a knack for taking the fledgling ideas of others and turning them into profitable inventions. Edison would eventually patent one of the earliest motion picture cameras, the kinetograph, and he used his invention to make short films. These were viewed with a companion projector, which he called a kinetoscope.

Born in 1847, Edison grew up in Port Huron, Michigan, evincing early on a strong curiosity about the world around him, and conducting his own experiments. In 1869, at age 22, he patented his first

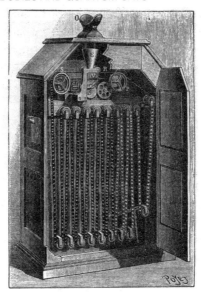

Edison's kinetoscope

invention, the electrographic vote recorder, a device that enabled legislators to register a vote for or against an issue by turning an electric switch to the left or right—with no risk of confusion from hanging chads. Thereafter he devoted his life to racking up more than 1000 inventions. He first started thinking about a motion picture camera in 1878, about the time he was granted a patent for the phonograph, a device that used a stylus to prick a pattern in a tinfoil cylinder in response to sound vibrations of, say, somebody's voice. A second needle would play the sound back as the patterned cylinder turned. Edison was keen to extend that technology to include combined moving pictures and sound, although it would be another 10 years before he discovered a means for doing so. "I am experimenting upon an instrument which does for the eye what the phonograph does for the ear, which is the recording and reproduction of things in motion," he wrote in 1888.

Impressed by Muybridge's photographic achievements, Edison met with him to discuss adding sound to the moving pictures.

Accounts differ as to whether Muybridge or Edison declined to collaborate, but Edison apparently thought the zoopraxiscope was impractical, because it used multiple cameras to achieve a series of images. Edison envisioned a camera capable of recording successive images in a single device, an approach that proved much more practical and cost-effective. "My plan was to synchronize the camera and the phonograph so as to record sounds when the pictures were made, and reproduce the two in harmony," he recalled later, in 1925. The basic concept was to use a cylinder similar to that in the phonograph, place it inside a camera, and then coat it with a light-sensitive material.

Serendipitously, it was about this time that George Eastman introduced the new celluloid film, which began to replace the old system of using light-sensitive plates and large bulky cameras, and led to the manufacture of the "Brownie" camera. This film was simply a strip of transparent plastic material (celluloid) with one side coated with several different chemical layers held together by gelatin. Edison ordered some of the new film cut into long strips. His assistant, William Dickson, developed a system to move the film past the lens when turned by a crank. This was the kinetograph. Every time a picture was taken, the cylinder rotated slightly, taking another picture. The crude film was then processed and run in slow motion through a viewer (the kinetoscope).

The first "film" Edison made was of Fred Ott, a worker in his laboratory in Newark, New Jersey. Ott acted out a sneeze on February 2, 1893. The sound of the sneeze was recorded on a phonograph, to be played back with the film, and the experiment proved a smashing success. Encouraged, Edison's team began producing movies in a studio in West Orange, dubbed "Black Maria" because of its resemblance to a police patrol wagon. By today's standards, it was extremely crude, with a hole in the ceiling to allow the sun to shine through and illuminate the stage. The entire building was constructed on a set of tracks to enable Edison's team to move it around and follow the course of the sun throughout the day. The films—which initially lasted only a few seconds—were shown on kinetoscopes placed in arcades around the country. Patrons could view short films of circus performers, dancers, or animals for a nickel; that is why the arcades came to be known as nickelodeons. Eventually, the team produced a 15-minute thriller,

"The Great Train Robbery," and went on to develop more than 2000 other short films.

Today's motion picture industry has far outstripped those humble technological beginnings, thanks to the advent of digital photography and exponential increases in computer power. The modern digital camera works on the same principle as a conventional camera, but instead of focusing light on a piece of film, it focuses light on an image sensor made of tiny light-sensitive diodes that convert light into electrical charges. It turns the fluctuating waves of light (analog data) into bits of digital computer data. Once it became possible to convert analog data into digital format, film technology took a huge leap forward, with vastly improved cinematography, sound, and highly sophisticated digital equipment capable of creating not only spectacular special effects, but also entire characters and fantastical worlds.

As recently as a decade ago, it would not have been possible to create Peter Jackson's Oscar-winning *Lord of the Rings* trilogy, which brought J. R. R. Tolkien's masterpiece vividly to life. It pushed the technology one step farther with a memorable half-human, half-computer-generated character: Gollum, a murderous former Hobbit whose lust for the One Ring (his "precious") is rivaled only by his penchant for sibilant speech and pathological misuse of first-person pronouns. Following in the footsteps of shadow casters like Villanova, Giambattista, and "Professor" Pepper, modern filmmakers are the new illusionists.

17

Radio Days

1895
Tesla demonstrates wireless radio

The story line of Roger Waters's concept album *Radio K.A.O.S.* (1987) centers on Benny, a Welsh coal miner and amateur ham radio operator. His twin brother Billy is wheelchair-bound, mentally challenged, and mute. But Billy has a secret, special skill: he can receive radio waves directly into his head, without the need for an antenna or receiver to translate the signals into sound waves. His unique ability enables him to pick up radio programming, police broadcasts, and even radio waves from outer space, which he believes are "messages from distant stars."

We can't see them, but radio waves are all around us, all the time. Like visible light, they are a part of the electromagnetic spectrum, only with much longer wavelengths. Radio waves can be as long as a football field, or as short as a football. This wide range makes them ideal for transmitting information, because different frequencies can be assigned to specific devices. We use radio waves not just in AM, FM, and police and fire department radios, but also in television, radar, cellular phones, baby monitors, remote control, and garage door openers, to name just a few. Each type of device has its own frequency range to avoid overlap.

Radio owes its existence to two earlier inventions: the telephone and telegraph. All three are closely related and exploit the flip sides of the electromagnetic coin. The essential components of any radio device are a battery and a metal wire. The battery acts as a simple transmitter, sending an electric current through the wire; the current creates a magnetic field around the wire. The higher the voltage of the current, the stronger the magnetic field, and the farther it can travel. Repeatedly connecting and disconnecting the battery causes changes in the magnetic field, giving rise to electric current in a second wire, which acts as a receiver. Connecting the battery creates the magnetic field and disconnecting collapses it; electrons flow in the second wire at those two moments. This process also emits radiation in the form of light waves in the radio region of the spectrum.

Exactly who invented the first wireless radio device has historically been a subject of much contention and legal action. The Italian inventor Guglielmo Marconi claimed to have invented his "black box"— essentially a spark transmitter with an antenna—at his home in Bologna, Italy, in December 1894. When Marconi first arrived in London in 1896 with his box, he entertained audiences with a "conjuring trick": making a bell ring on the other side of the room using radio waves instead of a cable. But the Serbian inventor Nikola Tesla had given the first public demonstration of radio communication in St. Louis, Missouri, the year before. Marconi's box was little more than a reproduction of Tesla's apparatus, which had been described in published articles. Marconi denied having read Tesla's articles, even though they had been quickly translated into several languages.

Thus began a years-long controversy, with both men vying to be recognized as the inventor and patent holder of what would prove to be a very lucrative technology. Although Tesla patented a wireless device as early as 1897, it was Marconi who became known as the "father of wireless communications," even sharing the 1909 Nobel Prize in physics for his contributions; Tesla was not a corecipient. Indeed, Marconi had aggressively developed and marketed the fledgling technology, aided by financial backers. But it was a grievous oversight by the prize selection committee to ignore Tesla's seminal contributions so completely.

Complicating the matter was the fact that at the time no one really understood what radio waves were, or how they worked—least of all Marconi. Even Tesla believed he had discovered a new kind of wave, instead of another part of the electromagnetic spectrum. At the time, the

transmissions seemed like magic to most people, leading many to conclude that wireless had supernatural properties and could be used for
communicating with the dead. Thomas Edison once attempted to build
a telephone to speak with the dead, and according to a popular urban
legend, Mary Baker Eddy was buried with a telephone—just in case she
had any valuable insights to offer from beyond the grave. We haven't
entirely lost that notion. It was the premise for the film *Frequency*
(2001), in which a son is able to communicate with his dead father,
30 years in the past, by means of an old ham radio unit, using the same
call letters, or assigned frequency, but in different time dimensions.

The first radio transmitters were called "spark coils" because they
created a continuous stream of sparks at high voltages. Their emissions
also spanned the entire radio spectrum, giving rise mostly to static. Still,
it was possible to use a standard telegraph key to switch the radio waves
on and off, thereby using the static to tap out messages in Morse code.
These signals were picked up and decoded at the receiving end.

Thanks to the wireless telegraph, the Arctic explorer Robert Peary
was able to use radiotelegraphy to tell the world about his discovery of
the North Pole in 1909. And in April 1912, as the lower compartments
of the supposedly unsinkable *Titanic* filled with water after the ship
was hit by an iceberg, the wireless telegraph operator sent out a series
of increasingly frantic Morse code messages, including what has
become the world's best-known distress signal: dot dot dot dash dash
dash dot dot dot (SOS).

Radiotelegraphy even helped apprehend an escaped murderer in
England. Early in 1910, Dr. Hawley Crippen chopped off his wife
Cora's head, tossed it into the English Channel, and buried the rest of
her body under the floorboards of his house. Crippen claimed that his
wife had run off with another man. The police believed him until
Crippen himself disappeared, along with his lover, Ethel LeNeve. The
chances of catching the pair seemed slim at best, but on July 22,
Scotland Yard received a wireless telegraph message from the captain
of the S.S. *Montrose*, bound for Quebec. The captain reported two
suspicious passengers who claimed to be father and son, but had
been seen holding hands on deck. They were Crippen and LeNeve, in
disguise. Technically, they were still on English soil as long as they
remained on the ship, so an inspector from Scotland Yard boarded

a faster ship and intercepted them before they could disembark in Montreal. Crippen was brought back to England to stand trial and was convicted of murder and hanged.

Once wireless telegraphy had been successfully demonstrated, attention turned to creating a wireless device capable of conveying speech and sounds, including music. The first step is identical to what happens with a telephone: sound waves are converted into an electric current through a transmitter. But instead of traveling along a wire, the encoded sound data hitches a ride on a radio wave. It does this by changing either the wave's height—also known as its amplitude—or its frequency, which is the number of times the wave oscillates per second.

Think of the unmodulated radio wave as a blank sheet of uniformly white paper, and the encoded sound wave as the ink that forms the printed words. The ink causes tiny variations in the background of the paper, and this is what the eye sees when it "reads." Old-fashioned long-playing records (LPs) rely on variations of the grooves in the vinyl to encode information, which is "decoded" by the tip of a phonograph needle. Compact discs (CDs) operate on much the same principle, but they encode data in "pits" instead of grooves, and the data are decoded by laser light. In fact, we get most of our sensory information from variations in background. In the case of radio, it's noise instead of paper and ink.

Unmodulated radio waves have the same height or frequency. Mixing in a data stream of electrons causes slight changes in either of those two properties. For example, AM (amplitude modulation) radio changes the amplitude, and FM (frequency modulation) radio changes the frequency. These changes are detected by the receiver, which then decodes the information. You can see these wave changes as they happen with instruments called oscilloscopes. They are often used in films and television dramas, such as *Enemy of the State*. When federal agents are monitoring telephone conversations from the home of their suspect du jour, whenever a voice speaks, the wavy green lines on the monitor change size and shape in response; this is similar to how a heart monitor works.

While Marconi was busy building a wireless telegraphy empire, Tesla—ever the visionary—had turned his attention to wireless radio broadcasts. He constructed the world's first AM radio broadcast tower

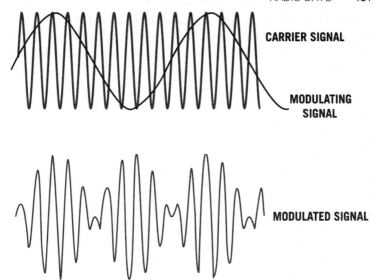

CARRIER SIGNAL

MODULATING SIGNAL

MODULATED SIGNAL

**Amplitude modulation
of a radio wave**

in Colorado Springs in 1899, the same year Marconi linked France and England with a wireless telegraph system across the English Channel. Tesla generated and transmitted wireless radio waves for miles. His transmitting antenna also served as a receiver, which he used to track thunderstorms. Encouraged by his experimental success in Colorado, he began constructing a giant radio station in Wardenclyffe, New York, in 1901, financed by J. P. Morgan, whose young daughter Anna had developed a schoolgirl crush on the handsome inventor. The transmitting tower was enormous, standing 187 feet high and capped by a 68-foot dome. It was intended to transmit radio signals without wires to any point on the globe.

But Tesla fell out with Morgan before the tower was completed, partly because the project was taking much longer to build than expected and was far over budget. The fact that Marconi succeeded with the first transatlantic wireless telegraph transmission in 1901 without such an enormous investment in equipment did nothing to bolster Tesla's case—even though Marconi accomplished his feat by using Tesla's patented apparatus. It was the beginning of the end for Tesla's scientific celebrity, and his wealth. He eventually signed over ownership of the Wardenclyffe Tower to creditors to pay his debts. The structure was razed in 1917 and sold for scrap.

For all his brilliance, Tesla was an eccentric, and his odd habits became even more pronounced as he aged. Like the "defective detective" Adrian Monk on television, Tesla suffered from obsessive-compulsive disorder, marked by a progressive germ phobia. He always wore white gloves and rarely shook hands. He never stayed in a hotel room or floor whose number was divisible by three; he feared pearl earrings worn by women; and at meals he insisted upon a copious supply of napkins, which he used to meticulously polish his silverware. In his last years, he made strange claims about death rays that could make entire armies vanish in seconds. He died virtually penniless on January 7, 1943, in the Hotel New Yorker, where he had lived for the past 10 years. Nine months later, the U.S. Supreme Patent Court determined that Tesla, not Marconi, should be considered the father of wireless transmission and radio, since there was nothing in Marconi's original patent that had not been previously registered by Tesla. Some historians have dubbed Tesla the "forgotten father of technology." Tesla himself said of the skeptics of his day, "The present is theirs. The future, for which I really worked, is mine."

Just how far ahead of his time was Tesla? In 1899, Tesla broached the possibility of interplanetary communication, earning the derision of his scientific peers. He was about 70 years too early. Millions of Americans were glued to their television sets on July 20, 1969, as the Apollo 11 astronauts Neil Armstrong and Edwin "Buzz" Aldrin became the first human beings to set foot on the moon. Shortly after disembarking from the lunar module, they received a phone call from President Richard Nixon back on Earth, who congratulated them on the achievement and declared, "This certainly has to be the most historic telephone call ever made."

There weren't any telephone poles or wires in the vast expanse of outer space, but there were radio waves. Nixon used a form of radio-telephony to make his call. It was an elaborate setup, comparable to sending a message by translating it from English into French into German, then back into French and English. The White House telephone converted the president's voice to electrical signals, which were transmitted over wires to Mission Control in Houston, Texas. These signals then traveled through space on a radio carrier wave that was transmitted to the moon via a communications satellite. The carrier wave was picked up by the antenna and receiver on the lunar module, which transmitted the signal to a receiver in the astronauts' space suits. This turned the electrical signal back into sound so that the astronauts could hear the president's voice. For the president to hear the astronauts speak, the whole process happened in reverse. All told, there was a delay of about 1.285 seconds each way between the transmitted signals.

Tesla's fascination with communicating with other planets stemmed from an unusual incident in 1899. One night in Colorado Springs, Tesla heard strange rhythmic sounds on his radio receiver, in such a regular pattern that he concluded it could only be an effort to communicate with Earth by alien beings from Venus or Mars. This wasn't as crazy as it sounds. Belief in extraterrestrial life was quite common at the time. Even today, the ongoing SETI project routinely analyzes radio signals collected from the skies for messages from intelligent alien life-forms—on the off-chance that Fox Mulder of *The X-Files* was right, and the truth really is "out there." Still, Tesla was roundly criticized for his claims, even though Marconi also claimed to have detected unusual pulsed signals that he believed to be extraterrestrial. And technically, they were: what Tesla and Marconi heard were radio waves emitted from distant celestial sources—planets, comets, stars, or galaxies—a phenomenon that would not be experimentally confirmed by scientists until 1932.

Today astronomers can monitor the skies with radio telescopes. Because radio wavelengths are so long, radio telescopes must use large dishes made of conducting metal to reflect radio waves to a focal point in order to achieve the same clarity and resolution as optical telescopes. In fact, astronomers often combine several smaller radio telescopes or receiving dishes into an array that acts as one very large instrument. (SETI gets its data from the Arecibo radio telescope, which fills an entire valley in Puerto Rico.) Data are collected from all the components and combined to enable scientists to infer the structure of radio wave sources in space. Like Billy in *Radio K.A.O.S.*, radio astronomers are quite literally detecting and decoding their own "messages from distant stars."

18

Mysterious Rays

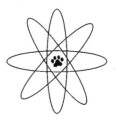

November 8, 1895
Röntgen discovers X-rays

"Only the gods see everything," a colleague warns Dr. James Xavier, the doomed antihero of the schlock-master Roger Corman's sci-fi cult film of 1963, *X: The Man with X-ray Eyes*. But Xavier, cursed with the hubris of a classical Greek tragic figure, thinks he has discovered a means of becoming like the gods. He has developed a serum that gives him "X-ray vision": the ability to see through most solid objects, like Superman—only he sports a white lab coat and pocket protector instead of a red cape and tights. The serum expands Xavier's vision beyond the range of visible light into the X-ray spectrum, enabling him to see through objects that would otherwise be impenetrable to the naked eye.

Like visible light, X rays are wavelike forms of electromagnetic energy (light) carried by tiny particles called photons. The only difference is the higher energy level of the individual photons, and the corresponding shorter wavelength of the rays, which makes them undetectable by the human eye. X-ray photons are produced in a fashion similar to visible photons. Orbitals are the paths electrons follow when circling around an atom, much like atomic-scale mini-planets circling the nucleus, which serves as a minuscule sun. Each orbital has a specific energy level. Electrons can jump between energy levels, and when they move from higher to lower levels, the excess energy is emitted as a photon. It's a bit like getting knocked down a rung on the corporate ladder. The sharper the drop, the higher the energy of the resulting light rays; in the case of X rays, the drop is steep, comparable to Donald Trump's being demoted to janitor.

So X-ray photons have energies that range from hundreds to thousands of times higher than those of visible photons. Because they are so energized, X-ray photons can pass through most materials. It all depends on the size of the atoms that make up the material; larger atoms absorb X-ray photons, but smaller atoms do not—the X rays pass right through. For instance, the soft tissue in the body is composed of smaller atoms and hence doesn't absorb X rays very well, whereas the calcium atoms in the bones are much larger and do absorb X rays.

This unusual effect lies at the heart of all X-ray machines, which essentially consist of a cathode and an anode inside a glass vacuum tube. The cathode is a heated filament of the kind you might find in a light bulb. Current passes through the filament to heat it, producing negatively charged electrons, which are then drawn across the tube by the positively charged anode, usually made of tungsten. The electrons are highly energized and move with a great deal of force. Whenever a speeding electron collides with a tungsten atom, it knocks loose an electron from one of the atom's lower orbitals. In response, an electron in a higher orbital must drop to a lower one to fill the gap, and releases its extra energy in the form of a photon. There is a big drop in energy, so the photon has a high energy level, in the X-ray range of the electromagnetic spectrum. A camera on the other side of the patient records the patterns of X-ray light passing through the patient's body. It's the same basic technology as in an ordinary camera, but X-ray light, rather than visible light, sets off the chemical reaction on the photographic plate.

We owe our knowledge of X rays and their unique properties to their accidental discovery by German-born physicist Wilhelm Röntgen. The only child of a cloth merchant and manufacturer, Röntgen moved with his family to the Netherlands when he was three. By the time he completed his secondary education at 17, he had evinced a keen interest in practical science. His early scientific career proved a bit rocky. Röntgen was expelled from the Utrecht Technical School, albeit unfairly, for a prank committed by another student: drawing an unflattering caricature of the physics teacher on the schoolroom's decorative stove screen. He nonetheless managed to enter the University of Utrecht in 1865 to study physics, and later moved to the Polytechnic at Zurich, where he studied mechanical engineering. He earned his PhD from the University of Utrecht and eventually obtained an academic position at the University of Würzburg, where he began conducting the experiments that would make him famous.

Röntgen was studying light emissions resulting from electrical current, using so-called "Crooke's tubes." These were glass bulbs with positive and negative electrodes, evacuated of air, which displayed a fluorescent glow when a high voltage current was passed through them. He was particularly interested in cathode rays, now known as electrons. In order to better observe the emissions, he enclosed the discharge tube in a sealed black carton to exclude all light, and worked in a darkened room.

On November 8, 1895, Röntgen noticed that the green light emitted by a Crooke's tube caused a fluorescent screen 9 feet away to glow. This alone was not unusual, since any fluorescent material will glow in reaction to light. But the tube was surrounded by heavy black cardboard that should have blocked the radiation. He concluded that the fluorescence was caused by invisible rays, which somehow penetrated the opaque black paper wrapped around the tube. While holding various materials between the tube and the screen to test these new rays, he saw the bones of his hand clearly delineated on the fluorescent screen within an outline of the surrounding flesh. He realized that this new type of ray was capable of passing through not just cardboard, but most substances, including the soft tissues of the body, while leaving bones and metals visible. One of the earliest photographic plates from his experiments was a film of his wife Bertha's hand, with her wedding ring as visible as the bones in her hand.

On December 28, Röntgen submitted a paper, "On a New Kind of Ray," to the Würzburg Physico-Medical Society. One month later, he made his first public presentation before the same society, following his lecture with a demonstration: he made a plate of the hand of an attending anatomist. He called his discovery "X" rays because they were a previously unknown type of radiation, and the name stuck. But initially, many people called them "Röntgen rays." The rays became an overnight scientific sensation. In the United States, Thomas Edison was among those eager to exploit the breakthrough; he developed a handheld fluoroscope, although he failed to make a commercial "X-ray lamp" for domestic use. An apparatus for producing X rays was soon widely available, and studios opened around the country to take "bone portraits," further increasing public interest. Poems about X rays appeared in popular journals. "I hear they'll go/Thro' cloak and gown—and even stays/Those naughty, naughty Röntgen rays," one would-be poet trilled in the April 1896 issue of the *Electrical Review*. Metaphorical uses of the rays popped up in political cartoons, short stories, and advertising. Detectives touted the use of Röntgen devices in following unfaithful spouses, and lead underwear was manufactured to foil attempts at peeking with "X-ray glasses."

Unlike many historic breakthroughs in basic science, which often don't find their way into practical applications for decades, X rays emerged from the laboratory and into widespread use in a startlingly brief time: within a year of Röntgen's announcement of his discovery, X rays were already being used for medical diagnosis and therapy; the practitioners were known as radiologists. By February 1896, X rays had found their first clinical use in the United States—in Dartmouth, Massachusetts, when Edwin Brant Frost produced a plate of a patient's fractured arm for his brother, a local doctor. In Dundee, Scotland, researchers used X rays to locate a bullet lodged inside the skull of a man who had accidentally shot himself several years earlier and miraculously survived.

Some experimentalists began using the rays to treat disease. Since the early nineteenth century, electrotherapy had proved popular for the temporary relief of both real and imagined pain. The same apparatus could generate X rays. Only a few days after the announcement of Röntgen's discovery, an electrotherapist based in Chicago irradiated a woman who had a recurrent cancer of the breast, and by the end of the year, several other researchers had noted the palliative effects of the rays on cancers. Others found remarkable results in the treatment of surface lesions and skin problems. X rays even found cosmetic uses in depilatory clinics in the United States and France, and they were used in shoe stores to measure foot sizes.

Röntgen was awarded the very first Nobel Prize in physics in 1901 for his work, and he donated the prize money to his university. When asked what his thoughts were at the moment of his discovery, he replied, "I didn't think, I investigated." He never sought honors or financial profits, and he rejected a title that would have given him entry into the German nobility. He also refused to take out any patents on X rays, so that the world could freely benefit from his work. His altruism came at considerable personal cost: at the time of his death in 1923, Röntgen was nearly bankrupt from the inflation that erupted in the wake of World War I. Yet he still fared better than some of his colleagues who devoted themselves to X-ray research.

In Corman's film, Xavier finds that X-ray vision goes far beyond the adolescent novelty of peeping through the clothing of comely young women; his gift quickly becomes a nightmare and spirals out of control, giving rise to horrifying visions, until at last he looks directly into the cosmic eye of an all-knowing god. He ends up gouging out his own eyes, like Oedipus, driven mad by seeing too much—things human eyes were never meant to see. X rays likewise proved to be a double-edged sword, with unforeseen consequences only slightly less dramatic.

In the early days following their discovery, X rays were widely considered to be harmless. Radiologists thought nothing of daily exposure to the rays, even though a dean at Vanderbilt University lost all his hair after sitting for a radiograph of his skull in February 1896. A Scottish researcher who routinely used a fluoroscope to test the quality of X rays by holding his hand between the tube and the fluorescent screen developed tumors and eventually lost both his hands. Other reported problems included redness of the skin, numbness, infection, and severe pain, and by the end of 1896 at least one scientist, Elihu Thomson, had concluded that X rays were the cause. But this view did not become widespread until 1904, when Edison's assistant, Clarence Dally, died from exposure, having carefully documented for posterity the burns, serial amputations, and diseased lymph nodes leading up to his death. Early efforts at protection were developed—lead screens, heavy aprons, and metal helmets—but by then it was too late for many of the early radiology pioneers.

Radiologists experienced the adverse effects because X rays are a form of ionizing radiation. Ions are electrically charged atoms, a by-product of an X ray knocking electrons off atoms. The resulting free electrons then collide with other atoms to create even more ions. This is dangerous because an ion's electrical charge can lead to unnatural chemical reactions inside cells, particularly at the higher energy levels of X rays. It can break DNA chains, causing the cell to either die or develop a mutation and become cancerous, which can then spread. And if the mutation occurs in a sperm or an egg, the result can be birth defects; this is why pregnant women should never be subjected to X rays.

With protective measures now firmly in place, X rays have ultimately proved to have more benefits than harmful consequences. It just so happens that the free electrons produced by X rays can also generate photons without hitting an atom. An atom's nucleus has magnetic properties that may attract a speeding electron just enough to alter its course, much like a comet whipping around the sun. This "braking action" causes the electron to emit excess energy as an X-ray photon. This is the effect exploited in the phenomenon of synchrotron radiation. When physicists began operating the first particle accelerators, they discovered that an accelerator's magnetic field would cause the electrons to move in large spirals around magnetic field lines of force. These particles can be accelerated to high energies near the speed of light, producing a beam of X rays as thin as a human hair, and 1 trillion times brighter than the beam produced by a hospital X-ray machine. It's a concept similar to the laser, but on a much larger scale.

Why would scientists deliberately make such a powerfully concentrated X-ray beam, knowing the potential damage it can cause? They do this to further our understanding of matter. We saw in Chapters 3 and 4 how visible light can be manipulated to let us "see" objects too small and objects too far away for the naked eye to detect. But atoms have dimensions of about 1/10 nanometer (1 billionth of a meter) and hence can't be seen with conventional microscopes using visible light. A much shorter wavelength is needed, and X rays fit the bill perfectly, as they are able to penetrate the deep atomic structure of all kinds of different materials down to the level of individual atoms.

This is important because atomic structure is what gives a material—whether it's a silicon chip, a spinach leaf, or a protein molecule—its unique nature. And the more we know about what that material can do, the more we can put it to good use to develop new drugs, tougher building composites, faster computer circuits, or tiny nanorobots for medical imaging from inside the human body, to cite just a few examples. So the high-energy X rays found in synchrotron radiation are giving scientists the ability to see things in nature previously invisible to the naked eye—though perhaps not yet the cosmic eye of the gods.

19

A Thousand Points of Light

October 1897
Discovery of the electron

In the film *Pleasantville* (1998), David is a high school misfit who longs to escape grim reality. His wish is unexpectedly granted. Thanks to a magical remote control provided by a rather creepy television repairman, he and his sister, Jennifer, are beamed into the monochromatic, decidedly G-rated realm of the fictional town of Pleasantville, and forced to take on the roles of two wholesome youngsters, Bud and Mary Sue, in the 1950s. They find themselves in a world of white picket fences, freshly baked cookies, and enormous breakfasts loaded with carbohydrates and fat. The high school basketball team never loses, married couples sleep in separate twin beds, and the local lovers' lane is a site for chaste hand-holding and pecks on the cheek. It's a town that could exist only in the pixel-painted world of television.

Television ("long-distance sight") is yet another technology that relies on the radio frequency range in the electromagnetic spectrum, just like the wireless telegraph and radio, both of which emerged in the late 1890s. We saw in Chapters 13 and 17 how sound can be turned into an electrical current and transmitted via radio waves. (Cable television applies the same basic principle but transmits the signals over underground cables to reduce interference; that's why we usually get better reception with cable systems.) These encoded radio waves are then picked up by a receiver and "decoded" back into sound.

That same phenomenon is exploited in TV broadcasting, which adds images to the transmission of sound. In Roald Dahl's novel *Charlie and the Chocolate Factory,* one of the children invited to tour the Wonka factory is Mike TeeVee, a television addict who is enthralled by the prototype for "Wonka Vision." He begs to be the first person to be transmitted over a TV signal: he is split up into millions of tiny pieces and sent whizzing through the air to the TV receiver, where he is put back together again. Alas, he ends up significantly smaller when he reemerges—the same size he would be on a TV screen. It's a clever twist on what actually happens during a TV broadcast. Images are scanned electronically, and break into thousands of tiny points of light, which are then grafted onto a radio wave for transmission. The image and the sound are sent separately, over different frequencies, and the TV set acts as a receiver, reassembling the scattered information to produce synchronized picture and sound.

Two critical elements are needed to make this possible: a cathode-ray tube (CRT), and tiny elementary particles called electrons. Traveling science lecturers in the mid-nineteenth century delighted audiences with a device that could be considered the ancestor of the neon sign. They used a glass tube with wires embedded in opposite ends, administered a high voltage current, and pumped out most of the air to create a vacuum. Known as a "Crooke's tube," after the British physicist Sir William Crooke, this device was the precursor of the modern CRT. When a current was applied, the tube's interior would glow in striking fluorescent patterns. Scientists theorized that the glow was produced by some kind of rays. But were these rays similar to light, which was then believed to be waves that traveled through a hypothetical medium called the ether, or aether? Or were they some kind of electrified particle?

Experiment after experiment only added to the confusion. For instance, when physicists moved a magnet near the glass, they found that they could maneuver the rays at will, indicating that the rays were electrically charged particles. But the rays were not deflected by an electric field. This confounded scientists. Electricity and magnetism had been found to be closely related, each giving rise to the other, so both should be able to deflect an electrically charged particle. And a particle would travel in a straight line, not diffuse outward like a wave. Yet when a thin metal foil was placed in the path of the rays, the glass still glowed, as though the rays had slipped through the foil. This indicated that cathode rays were waves. Other experiments showed that the rays carried a negative charge, and that the ratio of their mass to their charge was more than 1000 times smaller than the ratio of the smallest charged atom. If they were particles, they would have to be very tiny indeed.

It took a British professor in the Cavendish Laboratory at Cambridge University to solve the conundrum. The son of a bookseller, Joseph John (J. J.) Thomson showed an interest in science at a very young age, and when he was 14, his parents encouraged him to become an engineer. After qualifying as an engineer, Thomson began to study physics and mathematics, and in 1876 he received a scholarship to study at Cambridge University's Trinity College, where he would remain for the rest of his life. He earned a degree in mathematics in 1880 and joined the Cavendish laboratory, keen to conduct experimental research. And his attention soon shifted to those mysterious cathode rays.

First, Thomson tried to separate the negative charge from the cathode rays by bending them with a magnet. He failed; they were inseparable. Second, he set out to discover why the rays were affected by a magnetic field, but not by an electric field. It was known that a charged particle will be deflected by an electric field, but not if it is surrounded by a conducting material. He reasoned that the traces of gas still in the tube were acting as a conductor. So he extracted all the remaining gas from the tube, and lo and behold, when the gas wasn't present, the cathode rays did bend when an electric field was applied.

From this, Thomson concluded that cathode rays were "charges of electricity carried by particles of matter." He called these charged particles "corpuscles"; they wouldn't be called electrons until 1897. At a lecture at the Royal Institute, he also announced that these corpuscles were the constituents of the atom, then believed to be the smallest form of matter. His speculations met with considerable skepticism from his colleagues. One distinguished physicist who attended that famous lecture admitted years later that he believed Thomson had been "pulling their legs." Thomson thought electrons were the only constituents of the atom; we know today that this idea was false. Electrons are only the most common members of an entire family of fundamental subatomic particles, each with its own unique signature and properties. But without electrons, there would be no such thing as television.

Most early attempts to demonstrate the principle of television involved the use of spinning disks and mirrors to transmit a moving image through the air—a concept first proposed in the 1880s. For example, on December 2, 1922, a French engineer at the Sorbonne, in France, Edouard Belin, demonstrated a mechanical scanning device that took flashes of light and directed them at a selenium element connected to an electronic device that produced sound waves. (Belin made the first telephoto transmissions—from Paris to Lyon to Bordeaux and back to Paris—in 1907.) These sound waves could be received in another location and translated back into flashes of light on a mirror. The process was similar in concept to the "photophone" invented by Alexander Graham Bell in 1880, the precursor of modern fiber-optic telecommunications. But the setup was clumsy and not very robust, since the mechanical parts routinely broke down. The real turning point came in 1927, thanks to Philo Farnsworth, a farm boy turned entrepreneur.

Farnsworth was born in 1906 in Utah. He was expected to become a concert violinist, but he had other ideas. At age 12, he built an electric motor to run the first electric washing machine his family ever owned. He conceived the idea for television while he was still in high school, and researched picture transmission while attending Brigham Young University. After graduating, he cofounded Crocker Research Laboratories, later renamed Farnsworth Radio and Television Corporation.

Farnsworth proved that it was possible to transmit an image without using any mechanics whatsoever. He replaced spinning disks and mirrors with the electron itself, so small and light that it could be deflected back and forth within a vacuum tube tens of thousands of times per second. He was the first to form and manipulate a focused electron beam, and he also figured out how to turn an optical image into a stream of electrons. The first step is similar to what happens in photography: an image is focused by a lens onto a surface coated with light-sensitive chemicals. (Modern digital technology uses photosensitive diodes.) When hit by the reflected light, the chemicals emit an array of electrons in a corresponding pattern—the "electrical image"—that can then be turned into a fluctuating current. The voltage varies according to the brightness of the image: the brighter the image, the higher the voltage. Thus transformed, the image can easily be transmitted over a radio wave. Using a scanning electron tube he designed himself, Farnsworth transmitted the first TV signal on September 7, 1927, by wholly electronic means. It was merely 60 horizontal lines, more like the annoying broadcast test pattern than the complex moving images and sound of modern television. But it was a critical breakthrough.

A Russian-born American inventor working for Westinghouse is often cited as the "inventor" of television. It's true that Vladimir Zworykin received a patent in 1923 for a device called an iconoscope— essentially a primitive television camera. But he never demonstrated a working prototype. Farnsworth was the first to achieve a transmitted picture. He received a patent for his scanning tube in 1930. Zworykin didn't duplicate the achievement until 1934, and didn't receive a patent for his own version of a scanning tube until 1938. By then he was working for RCA, whose president, David Sarnoff, vowed that RCA would control TV the way it then controlled radio: "RCA earns royalties; it doesn't pay them," he declared. The company launched a $50 million legal battle against Farnsworth. The crux of the case came when Farnsworth's high school science teacher traveled all the way to Washington, D.C., and testified that at 14, Farnsworth had shared his rudimentary concept of a TV scanning tube with his teacher. The U.S. Patent Court ruled in favor of Farnsworth. RCA finally capitulated in 1939 and bought a license for the patents.

Television has come a long way since the first black-and-white TV broadcasts were made in the late 1930s; by the 1950s, TV sets were becoming commonplace. But the basic technology remains the same. Open up your TV set, and one of the first things you'll see is a bulky glass contraption: the CRT. Think of the two ends of a battery: one (the anode) is positively charged, and the other (the cathode) is negatively charged. Inside the CRT, the cathode is a heated piece of wire called a filament, similar to that found in typical lightbulbs, which naturally emits a stream of electrons. All the air has been pumped out of the tube to create a vacuum. And since electrons are negatively charged, they are attracted to the positively charged anode at the other end of the tube. So they fly through the vacuum in the tube in a tight, high-speed beam and strike the inside of the flat screen at the other end of the TV.

That screen is coated with a phosphor. Phosphors are materials that emit visible light when exposed to some kind of radiation, such as a beam of electrons. There are thousands of kinds of phosphors, and each emits light of a specific color. When an electron beam strikes the phosphor, the screen glows. Why does the beam create a picture, instead of simply landing as a tiny glowing dot smack in the center of the screen? Remember that nineteenth-century physicists were able to "push" electrons around by using a magnetic field. Inside your TV are magnetic "steering" coils: copper windings that create magnetic fields inside the tube to move the electron beam across and down the screen, line by line from left to right, to gradually "paint" an entire image out of tiny phosphor dots.

Black-and-white screens use only one type of phosphor, which glows white. The intensity of the beam varies to create different shades of black, gray, and white across the screen. For a color screen, there are three phosphors arranged as dots (called pixels) that emit red, green, and blue light—the three primary colors of light most easily absorbed by the human eye. (These should not be confused with the pigment primary colors routinely taught in art schools.) There are also three separate electron beams based on the incoming signal to illuminate the three colors. (Color images are sent the same way as black-and-white images used to be transmitted, but there is a separate signal for each of the three colors used.) Close to the phosphor coating is something called a shadow mask: a thin piece of metal punctured with tiny holes that line up perfectly with the phosphor dots on the screen. When a red dot is needed, the TV fires the red electron beam at the red phosphor; ditto for the green and blue phosphor dots. The rest is basic color theory: to make a white dot, all three beams fire simultaneously and the colors mix together to make white. For black, all three are turned off simultaneously as they scan past that portion of the screen. All other colors are just various combinations of the primary colors: red, green, and blue.

There's one other factor that makes television possible: an odd little quirk of human perception. The print in newspapers and magazines is really made up of tiny dots. We perceive them as a single image because those dots are spaced so closely together. The brain can also integrate movement. The flip books we played with as children or a reel of film essentially divides a motion sequence into a series of still shots. When those images are "replayed" in rapid succession, the brain reassembles them into a single moving scene. And it does this in fractions of a second. The lines "painted" on the inside of a TV screen by the phosphor dots are spaced so close together that the brain integrates them into a single moving image.

In the film *Batman Forever*, the Riddler is a demented inventor who creates a three-dimensional holographic TV system with an "interesting side effect": it sucks out the viewer's brain energy and channels it to the Riddler himself, making him smarter at the expense of the viewing population. Television (aka the "idiot box") is often blamed for the "dumbing down" of America. Even Farnsworth wasn't a fan of the technology he helped create, once telling his family, "There's nothing on it worthwhile, and we're not going to watch it in this household, and I don't want it in your intellectual diet." It's almost as if he foresaw daytime television, the Home Shopping Network, and the current glut of reality game shows. But technology is neither intrinsically good nor innately evil; this depends on how it's used.

The presence of David and Jennifer in Pleasantville slowly begins to alter the town's reality, but it ultimately proves to be a positive change. (As Jennifer puts it, "Nobody is happy in a poodle skirt and five pounds of underwear.") The townspeople start to move beyond their assigned stock character roles and behave in new, unexpected, and very human ways. Splashes of color begin to appear, until ultimately the entire town bursts into a gorgeous panorama of Technicolor hues. Library books have content instead of just blank pages, and Main Street finally leads somewhere, instead of simply going around in circles. The town is no longer isolated. At its best, that's what television does: it brings together a thousand tiny points of light to create a coherent moving image that can be viewed by people all over the world. Television is the great connector.

20

Quantum Leap

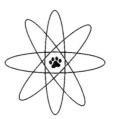

October 1900
Planck introduces quanta

The highly respectable scientist Henry Jekyll has a strange relation-ship with his younger assistant, Edward Hyde, in Robert Louis Stevenson's classic nineteenth-century "shilling shocker" *Dr. Jekyll and Mr. Hyde.* The two are in fact the same man. Jekyll was "wild" in his youth, before settling into his comfortable middle-aged life. But he is haunted by the darker, roguish side of himself that he has kept sub-merged for so many years. So he concocts a chemical potion that will, he hopes, separate the "good" and "evil" aspects of his personality. What emerges is Edward Hyde, a "pale and dwarfish" alter ego who thinks nothing of trampling a little girl or beating an older gentleman to death with his cane. Hyde is as base and lacking in moral qualms as his counterpart is noble and virtuous. Jekyll becomes, quite literally, a man with two faces.

Stevenson's tragic tale of a man wrestling with his darker impulses challenged Victorian assumptions about individual identity and gave birth to the modern notion of the dual personality. A fascination with duality still pervades the popular imagination. Surely it's no coinci-dence that so many of our favorite superheroes have two faces. Super-man spends his days as the unassuming newspaper reporter Clark Kent, and Spiderman pays the bills as the newspaper photographer Peter Parker. Batman's alter ego, Bruce Wayne, has the good fortune to be independently wealthy, but nonetheless is driven by his personal demons to fight crime every night in a cape and tights. Even the despicable Mr. Hyde is recast as an unlikely (and somewhat reluctant) hero in Alan Moore's graphic novel series, *The League of Extraordinary Gentlemen*. Jekyll–Hyde joins forces with other nineteenth-century fictional characters—Allan Quatermain, Captain Nemo, the Invisible Man, and *Dracula's* Mina Murray—to foil the nefarious schemes of evildoers intent on world domination.

So it shouldn't be surprising that duality also pops up in physics—minus the moral overtones. Light also has two faces, and its dual nature long challenged scientific assumptions about how atoms behave, ultimately leading to the quantum revolution in physics. Since Pythagoras in the fifth century BC, scientists had been flip-flopping back and forth as to whether light was a particle or wave. Pythagoras was staunchly "pro-particle." Two hundred years later, when Aristotle suggested that light travels as a wave, he was roundly ridiculed by his contemporaries. They reasoned that light clearly traveled in a straight line and bounced off a mirror, and surely this kind of behavior placed it firmly in the particle category. But light also diffused outward, like a wave, and different beams of light could cross paths and mix. By the mid-seventeenth century, Aristotle's notion of light waves had gained considerable acceptance among scientists, who based their concept of light on other wave phenomena found in nature, like sound or water waves.

Isaac Newton started experimenting with light while he was still a young man. One of the first things he discovered was that light is a mixture of colors. Prisms had been around for centuries—as soon as broken glass was available, people noticed that colors appeared wherever two refracting surfaces formed a sharp edge. No one knew where the colors came from, but most assumed that the triangular prism was somehow creating them. In 1666, Newton performed what he called his *experimentum crucis* using a pair of prisms. He darkened his study and made a hole in the window shutter to let in a single beam of sunlight, then believed to be the purest form of light, with no intrinsic color. The sunbeam passed through the first prism, which separated it into colors. Then Newton rotated the first prism in his hand, directing first blue light, then red light through the second prism, and discovered that it couldn't be separated any further. So white light was not

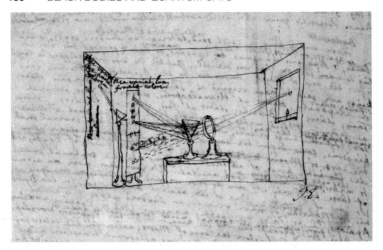

Newton's *experimentum crucis*

pure; it was a mix of various colors. And a prism didn't create colors—
it separated them out of the apparent white light.

Newton also concluded that light was made up of streams of particles
that he called "corpuscles." His primary argument against wave theory
harked back to Pythagoras: unlike sound waves, light does not turn cor-
ners; it always travels in a straight line. In 1672, Newton was persuaded
to publish his conclusions about light in the *Philosophical Transactions*
of the Royal Society, to which he had just been elected. But his ideas
didn't meet with the hoped-for reaction. Instead, he found himself
embroiled in a contentious and sometimes bitter four-year debate with
colleagues who clung to the wave theory of light. Always a bit prickly
and prideful, Newton refused to publish anything further on the subject. His last major work, *Opticks*, wasn't published until 1704.

Most of Newton's contemporaries—including Robert Hooke, the
author of *Micrographia;* and the Dutch mathematician and astronomer
Christiaan Huygens—favored a wave theory of light. And a pivotal
experiment in 1801 by the British physicist Thomas Young appeared to
clinch the matter. He shone light through two small slits in a screen,
and found that it formed an interference pattern (alternating light and
dark bands) on the other side. To Young, this indicated that light acts

like a water wave rippling outward when a stone is dropped in a pond. If the light encounters a barrier with a couple of openings, it splits into two different parts as it passes through the barrier. When the light recombines on the other side, the two parts either add together to produce a band that is twice as bright, or cancel each other out to produce a dark band. More than 50 years later, James Clerk Maxwell concluded that visible light was simply a form of electromagnetic radiation, which also includes radio waves, microwaves, infrared (heat), ultraviolet light, X rays, and gamma rays. So light is made up of moving electric and magnetic fields. Maxwell based his set of electromagnetic equations on the assumption that light was indeed a wave.

Both Newton and Maxwell were right, although neither could have predicted the revolution that was to come. The man who would lead twentieth-century physics into uncharted waters was a German physicist, Max Planck. Hailing from a long line of academics, Planck excelled more in music than physics and mathematics during his early

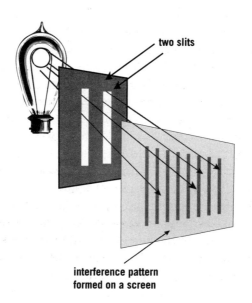

two slits

**interference pattern
formed on a screen**

**Shining a light through a double slit produces a
wave-like interference pattern**

education. A gifted pianist, he briefly considered a career in music, but when he asked a musician about the prospect, he was told that if he had to ask the question, he should probably study something else. So he turned to physics. His physics professors were hardly more encouraging; one told the young Planck in the early 1870s that physics was essentially a complete science with little potential for further development. But Planck was drawn by the seemingly absolute nature of the laws of physics and how they could be used to describe real-world phenomena. He was especially taken with the absolutism (or so he thought) of the second law of thermodynamics, according to which a closed system always loses a certain amount of energy from friction and other forces, thanks to entropy.

In a sense, modern quantum mechanics grew out of thermodynamics, specifically a nineteenth-century equation that related the temperature of an object to the total amount of radiation energy it emits. For example, if an object's temperature is doubled, the amount of radiation it emits will increase 16-fold. By 1900, physicists had figured out that the heat energy of a substance was the result of its atoms jiggling around; this movement emits ripples of radiation. The hotter the atoms get, the faster they jiggle, and the higher the energy (frequency) of the radiation produced. The trouble started when scientists tried to calculate the amount of radiation that would be emitted by an empty cavity: a hypothetical perfect emitter and absorber, which they called a "blackbody."

Imagine an enclosed box with a tiny hole in one side, whose walls are maintained at a uniform fixed temperature, like an oven. The jiggling atoms that make up the blackbody would emit radiation (as heat energy) through the hole, and the temperature of the walls would be the same as the temperature of the radiation inside. Thermodynamics tells us that heat flows from a hotter body to a colder one. If the box's walls had more energy than the enclosed radiation, energy would transfer from the walls to the radiation until a state of equilibrium was reached. The same process would happen in reverse if the radiation had more energy than the walls.

But what happens as the blackbody heats up? In 1792, Thomas Wedgwood (son of the famous porcelain manufacturer Josiah Wedgwood) noticed that raw materials changed colors as he raised the temperature inside his ovens. In 1900 German scientists conducted experiments to determine how much radiation came off objects at various temperatures. They expected that as the temperature rose, the amount of emitted radiation would rise accordingly. They found just the opposite: the amount of radiation didn't get bigger and bigger as the wavelengths grew shorter; instead, it hit a threshold and began getting smaller and smaller again. Planck's goal was to devise a theory that would explain these strange results.

Physicists could easily calculate the energy carried by an individual electromagnetic wave using Maxwell's equations, but the theoretical blackbody contains waves of varying frequencies ranging along the spectrum, from zero to infinity. There are a fixed number of atoms that make up the box itself, so it's relatively easy to figure out how to divide the total amount of energy between them. It's not that simple to figure out how to divide a fixed amount of energy among the infinite mix of radiation waves inside the box. You can't split a finite amount of energy equally among an infinite number of wave frequencies.

Planck—then a physics professor at the University of Berlin—decided that each atom making up the walls of the cavity could only absorb or emit radiation energy proportional to its frequency, or else it carried none at all. And that frequency could increase only in multiples of this basic amount. He called those steps "quanta," from a Latin word meaning "how much," because he didn't know exactly how much energy was in a quantum. The notion struck most physicists at the time as bizarre, because nature doesn't seem to move in jumps—certainly not on the macroscale. In *Zero: The Biography of a Dangerous Idea*, the science writer Charles Seife compares Planck's quanta to having people who are 5 feet tall and 6 feet tall with nothing in between, or cars that can travel at 30 or 40 miles per hour but never at 33 or 38 miles per hour. It seemed absurd. But it worked. It solved the issue of infinity by effectively placing a cap on how much the radiation frequency inside the box could increase. The minimum unit of energy could never exceed the total amount of energy in the heated box, so scientists could simply ignore all theoretical radiation above a certain frequency.

Planck has been described as a "reluctant revolutionary" by more than one science historian. There was nothing in his quiet, respectable background to indicate that he would one day change physics forever, and he could never have predicted the impact that quanta would have. It was a self-described act of desperation: "A theoretical interpretation had to be found at any price, no matter how high that might be," he later said, upon receiving the Nobel Prize in physics in 1918. And it did come at a price: Planck was ultimately forced to reject his belief that the second law of thermodynamics was absolute, and instead embrace the more controversial view that it was a statistical law of probability. Not all frequencies were radiated with equal probability. Instead, the probability decreased as the frequency of the radiation increased.

Planck thought his concept of quanta was just a mathematical "trick" to get theory to match experiment. After all, he was describing a light wave as if it were a particle. But it became increasingly obvious that his notion of little packets of energy could explain a lot of puzzling experimental results. The most notable was the fact that certain metals emit electrons when hit with a beam of light; this is known as the photoelectric effect. The German physicist Heinrich Hertz accidentally discovered the phenomenon in 1887, when he noticed that shining a beam of ultraviolet light onto a metal plate could cause it to shoot sparks.

It wasn't the emission that was surprising. Metals were known to be good conductors of electricity, because the electrons are more loosely attached to the atoms and could be dislodged by a sudden burst of incoming energy. What was puzzling was that different metals required bursts of different minimum frequencies of light for the electron emission to occur. For example, green light was sufficient to knock off electrons from a piece of sodium metal, but copper or aluminum required higher-energy light, in the ultraviolet (UV) range. Physicists also found that increasing the brightness of the light produced more electrons, without increasing their energy. And increasing the frequency of the light produced electrons with higher energies, but without increasing the number produced. (The frequency of a light wave describes how close together the peaks and valleys of the "ripples" will be. Brightness describes how high and low those peaks and valleys will be, a property known as amplitude; square the amplitude of a light wave, and the resulting number will be its brightness. The two properties are related, but distinct, since one can increase the frequency of light without increasing its brightness.)

In March 1905, Albert Einstein—still a lowly patent clerk in Switzerland—published a paper explaining this puzzling effect by extending Planck's quanta to light. (Planck had assumed that just the vibrations of the atoms were quantized.) Light, Einstein said, is a beam of quantized energy-carrying particles (later called photons), whose frequencies can increase only in specific amounts, much as an ATM machine can dispense cash only in multiples of $20. When that beam is directed at a metal, the photons collide with the atoms. If a photon's frequency is sufficient to knock off an electron, the collision produces the photoelectric effect. That's why different minimal frequencies were needed for different metals. Making the light brighter would bombard the metal's surface with even more photons, so naturally there would be a corresponding increase in the number of free electrons generated. Increasing the frequency of the light produces photons of higher energies, but each is still able to knock off only a single electron. That's why you see the same number of electrons produced, but at higher energies. As a particle, light carries energy proportional to the frequency of the wave; as a wave, it has a frequency proportional to the particle's energy. Einstein won the Nobel Prize in physics in 1921 for his work on the photoelectric effect.

Any lingering doubts were laid to rest in 1922, when an American physicist named Arthur Compton conducted an experiment providing even more evidence that light behaved like individual particles. Several physicists had noticed that if X rays were bounced off of certain crystals, the rays would reemerge from the collision with less energy than before, and the decrease in energy corresponded with the angle of reflection. To Compton, this could make sense only if X rays were made of photons that bounced off the electrons in the crystal atoms. He imagined a billiard ball effect: if one moving ball directly hits a stationary one, it transfers a chunk of its energy to the second ball, sufficient to overcome inertia and cause the second ball to move in the direction of the force. If it is only a glancing collision, rather than a direct hit, there won't be nearly as much energy transferred, and the second ball will move only slightly, if at all, in the direction of the force.

Physicists were forced to accept that—like Jekyll and Hyde or Superman and Clark Kent—light has a dual identity. It behaves like collections of tiny particles, and yet it has the wavelike properties of frequency and wavelength. And thereafter physics was forever divided into classical and quantum systems. Today, Maxwell's wave equations are used in studying the propagation of light as waves. Yet scientists must use the quantized particle approach for studying the interaction of light with matter.

Gazing into a mirror the first time he transforms himself into Hyde, Stevenson's Jekyll initially feels no repugnance, concluding, "This, too, was myself." But the situation soon spirals beyond his control, and Jekyll ultimately rejects his alter ego, as Einstein would later reject the quantum revolution he helped create. The notion of quanta, and the dual personality of light, turned out to have some very disturbing implications for the nature of physical reality. Within 20 years physicists would discover that not just light, but particles of matter, also had two faces. And the realm of tiny particles turned out to be governed by very different physical laws. But that's a story for another chapter.

21

It's All Relative

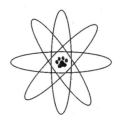

June 1905
Einstein introduces special relativity

Marty McFly, the teenage hero of the hugely popular film *Back to the Future* (1985), is a typical American high school student: he plays in a rock band, skateboards, and covets a crucial set of wheels so he and his girlfriend can ride around town in style. Alas, the "hip" gene appears to have bypassed the other members of his family. His alcoholic mother is clearly battling clinical depression, and his father is an obsequious, wet noodle of a nerd clad in polyester and horn-rimmed glasses. Marty's brother is a high school dropout who works at a fast-food restaurant, and his pudgy, fashion-challenged sister can't get a date to save her life.

Even Marty has his geeky side: he befriends an eccentric mad scientist, Doc, whose makeshift laboratory is crammed with bizarre inventions. After 20 years of labor, Doc has finally created a working time machine, housed in the shell of a pricey DeLorean automobile. In Hollywood, the course of scientific experimentation never runs smooth, so naturally Marty finds himself accidentally transported back to the 1950s, and hilarious high jinks ensue.

The notion of time travel is one of the most prevalent themes in science-fiction novels and films, at least since H. G. Wells published *The Time Machine* in the late nineteenth century. The film *Time after Time* transports a fictionalized Wells to San Francisco in the 1970s in pursuit of Jack the Ripper, who has escaped Victorian England in a time machine of Wells's devising. In the French film *The Visitors*, a medieval knight and his ill-kempt page find themselves trapped in modern-day France. Woody Allen's *Sleeper* awakens to his own version of a futuristic nightmare: public phone calls cost $1 million. Inspired in their turn, scientists have theorized about how time travel might one day be possible, by means of rotating black holes, wormholes, or cosmic strings. But it was special relativity—proposed in June 1905 by Albert Einstein, 10 years after the publication of *The Time Machine*—that first suggested a theoretical basis for this far-fetched stuff of science fiction.

Einstein's fascination with science began when he was four or five and first saw a magnetic compass. He was enthralled by the invisible force that caused the needle to always point north, and the instrument convinced him that there had to be "something behind things, something deeply hidden." He spent the rest of his life trying to decipher the arcane mysteries of the universe. Today, the name Einstein is synonymous with genius, but for years his parents thought their son was a bit "slow" because he spoke hesitantly and wasn't a stellar student. Einstein was just plain bored with the teaching methods of formal education, with its emphasis on memorization and blind obedience to an arbitrary authority. He preferred to study at home with books on math, physics, and philosophy. "It's almost a miracle that modern teaching methods have not yet entirely strangled the holy curiosity of inquiry," he later said. "For what this delicate little plant needs more than anything, besides stimulation, is freedom."

Einstein found both when he attended a local Swiss school in Aarau, having failed the entrance exam for the more prestigious Swiss Federal Institute of Technology. For the first time, he had teachers who gave him the freedom and latitude to pursue his own ideas, and he threw himself into studying the electromagnetism theories of James Clerk Maxwell, which were rarely taught at universities at that time. Then he studied physics at the Institute of Technology in Zurich, but he graduated with an undistinguished academic record and failed to obtain a university post teaching mathematics and physics. Instead, he worked as a patent clerk in Bern, doing theoretical physics on the side, and occasionally meeting with a group of friends to read and discuss books on science. They called themselves the "Olympia Academy."

At the beginning of 1905, Einstein recalled, "A storm broke loose in my mind." That storm might be more aptly called a hurricane: his ideas would irrevocably change the face of modern physics. Two fundamental concepts underlie special relativity. First, anything that happens in the universe involves both three-dimensional space and the fourth dimension of time: this is known as the space-time continuum. Second, according to special relativity we can't know that an event has occurred until information about that event reaches us. Light travels incredibly fast—669,600,000 miles per hour, to be precise—but there is still a delay before news of the event reaches the observer and enables him or her to react to the information. We often see this effect on the evening news: there is always a short delay between question and response when an anchor is speaking to a correspondent on the other side of the globe, and there is always a short delay before the visual image reaches us.

In fiction, a story may have a narrator with an individual perspective that colors his or her perception of events. Some stories have more than one narrator. For example, the classic film *Rashomon* relates the story of a man's murder and his wife's rape from four different and contradictory perspectives. Iain Pears uses a similar technique in his best-selling historical mystery *An Instance of the Fingerpost*. The book has three sections, each of which is narrated by a different character, who relates the same sequence of events in a slightly different way according to his or her unique point of view.

Special relativity extends this notion to the physical world. In physics, a frame of reference simply denotes where an observer happens to be standing; it's similar in concept to the narrative point of view in fiction. Einstein's equations revealed that there is no such thing as a universal, completely stationary frame of reference because every object in the universe is moving through time and space. Even if Marty McFly thinks he's standing still, the Earth itself is moving, and he is moving with it. And if Marty and Doc view the same event from different perspectives, each of them will see it in a different way. As Marty cruises in the DeLorean, while Doc stands in the parking lot, Doc is stationary and Marty is moving away from him. But Marty could assert that he is stationary and Doc is moving away from him—unless Marty is accelerating, in which case he will feel the counterforce to acceleration and conclude that he is indeed the person moving. Otherwise, mathematically, there is no distinction. Both statements are correct, according to Doc's and Marty's individual frames of reference.

That doesn't mean there aren't any absolutes. The laws of physics still apply equally in every individual frame of reference. This means that the speed of light is constant in all frames of reference. Maxwell's equations had indicated that the speed of light is constant, but until Einstein, most physicists believed that light would show the effects of motion. Yet experiment after experiment failed to turn up any evidence to support their belief. Maxwell's equations were correct; the interpretation was wrong. In reality, if Doc were to measure the speed of light, he would get the same answer whether he was at home in his laboratory, or cruising in the DeLorean.

This notion flies in the face of our daily experience: most physical objects add their speeds together. For instance, let's say Marty throws a rock at the bully Biff while riding by on his skateboard. Marty is traveling at 5 miles per hour, and the stone travels at 3 miles per hour. If Doc were standing nearby with a radar gun, he would clock the stone's speed at 8 miles per hour. If Marty were to clock it, the stone would be moving at 3 miles per hour, since he is already traveling at 5 miles per hour. But if he were to launch a beam of light from his moving platform, both Doc and Marty would measure its speed as 669,600,000 miles per hour. The cruising speed of 5 miles per hour would have no effect whatsoever.

Clearly something odd is happening. Einstein found that whenever an object with mass is in motion, its measured length will shrink in proportion to its speed; the faster it goes, the more its length will shrink. No object with mass can reach exactly the speed of light, but if it did, it would shrink to nothing. This is known as length contraction. Because time and space are linked, something similar happens with time: it slows down with motion, a phenomenon known as time dilation. None of this is apparent to the object that is moving; it is apparent only to an outside observer in a different frame of reference. As Marty speeds up in the DeLorean, everything in the car—the armrests, the seats, even Marty himself—is contracting in length, but because Marty is moving with the car, he doesn't notice any change. If

Doc, from his stationary vantage point, were able to observe and measure the car's contents while it was in motion, he would be able to see the contraction.

But light has no mass, so it can travel at exactly the speed of light, and length and time contract to nothing. That's why no matter what distance an outside observer measures for light to travel, over any given time frame, from the perspective of someone riding on a light beam, no distance would be covered at all, and everything would occur in an eternal zen-like "now."

The effects of length contraction and time dilation mean that there can be no such thing as two simultaneous events, when each is viewed from a different frame of reference. Space and time are not fixed. The distance between two points in space, and the time between two events, will depend upon the observer's point of view. Let's say that Biff and one of his cronies are lying in wait for Marty, still trapped in the 1950s. The crony positions himself as a lookout on one end of the block, at point B. Biff positions himself at the other end of the block, at point A:

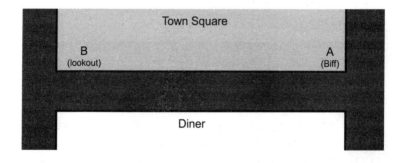

**Hill Valley's
town square**

Then Marty comes whizzing down the street on his skateboard, pulling a long banner behind him. A patron in the diner across the street notes how long it takes for the front end of Marty's skateboard to reach point A, and how long for the tail end of the banner to pass point B. From the patron's perspective, the lengths of the skateboard and banner are contracted. So the tail end of the banner will reach point B before the front of the skateboard reaches Biff at point A:

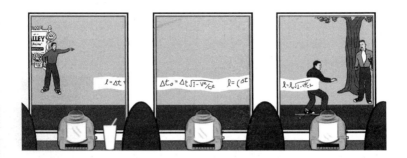

**Length contraction
from diner patron's point of view**

But from Marty's frame of reference, the length contraction occurs between Biff and his crony. So he would conclude that the front of his skateboard reaches Biff at point A well before the back of the banner reaches point B. Both conclusions would be correct.

**Length contraction
from Marty's point of view**

If you aren't developing a headache right about now, you haven't been paying close enough attention. Special relativity is a mind-bender. But what does all this abstract theory have to do with time travel? The link can be found in Einstein's most famous thought experiment, the "twin paradox." Since there are no twins in *Back to the*

Future, let's use the Olsen twins as an example. Mary Kate and Ashley share the same frame of reference while shooting a futuristic space-age film together on a Hollywood set. They synchronize their matching designer watches, and then Mary Kate blasts off in the set's spaceship—which turns out to be fully operational, and not just a cardboard-and-Styrofoam prop. She travels into space, while Ashley stays behind on earth. The twins are now in different frames of reference. This is where time dilation kicks in: were they able to view each other's watches, each would perceive the other's as moving more slowly.

But there's a twist: when Mary Kate returns to the set, she will have aged more slowly than Ashley because time passed more slowly for her than it did for her twin sister. Her watch will show that it took less time for her to go and return than Ashley's watch will show. According to what we've learned so far, when Mary Kate and Ashley once more share the same frame of reference, the effects of time dilation and length contraction disappear; their watches should be in sync. So what happened? At some point, Mary Kate turns around for the return trip. In so doing, she switches frames of reference, from one in which she is moving away from Ashley to one in which she is moving toward Ashley. That action breaks the symmetry between them, so that they are no longer in sync when Mary Kate returns to her sister's frame of reference.

Mary Kate gained only a few fractions of a second, but technically, she returned to the future. So it is conceivable that someone could travel farther than a few nanoseconds into the future by moving at faster and faster speeds. The bad news for aspiring time travelers is that it's well-nigh impossible to travel into the past. In order to do so, we would need to travel faster than the speed of light, and not even Doc's souped-up DeLorean could manage that. (For reasons known only to the filmmakers, the DeLorean jumps into a time-travel mode when it hits 88 miles per hour.) Light sets the cosmic speed limit: it is believed to be the highest speed at which an object without mass, such as a photon (a particle of light), can travel.

An object with mass, like Marty, can never reach the speed of light. That's because an object's mass (as measured by an outside observer) will increase as the speed increases. Marty will get heavier the faster he goes, and the amount of energy required to continue accelerating will increase proportionally. An object's mass becomes infinite at the speed of light, so it would require an infinite amount of energy to reach the full speed of light. Even if Marty could find a source of infinite energy, at some point, he would have to reach exactly the speed of light before he could surpass it. And thanks to that pesky length contraction, at the speed of light, an object contracts to nothing. Marty would shrink to a wafer-thin, two-dimensional version of himself, and possibly wink out of existence. It's unlikely that he would survive a theoretical transition to the other side of the speed-of-light barrier.

There are other, more philosophical complications. Within a few hours of his arrival in the 1950s, Marty accidentally encounters both his parents, and inadvertently keeps them from meeting. This puts his very existence into question unless he can correct the anomaly. It's a postmodern twist on the "grandfather paradox": if someone traveled back to a time before he was born and killed his own grandfather, then how could that person be alive to travel back in time and kill his own grandfather? If traveling into the past were possible, we would be able to rewrite history. But there is no guarantee that our revisionist "future" would be an improvement.

The science-fiction author Connie Willis found a clever solution to this paradox in two novels about time travel: *The Doomsday Book* (which won a Hugo award) and *To Say Nothing of the Dog*, both set in the late twenty-first century. In Willis's fictional world, time travel is not only possible but almost commonplace. So-called "temporal physicists" have invented a system called "the Net," which transports researchers into the past—historians, mostly, eager to study their subject firsthand. But there are still strict physical laws governing its operation. For instance, travelers are automatically prohibited from landing at any point in space and time where they could affect the course of history. Nor can physical objects be transported from the past into the future, foiling the plans of would-be scavengers. The system corrects itself to guard against such so-called "incongruities."

In the real world, physicists have developed another way of explaining such paradoxes: the notion of parallel universes. According to one theory—and it is still very much in the realm of theory—there may be an infinite number of universes. When Marty prevents his parents from meeting, he does so in only one universe, which is no longer the universe he exists in. Marty ultimately corrects the life-threatening paradox and makes sure his parents get together, but he still irrevocably alters the future. He returns to find his father a confident, successful novelist; his mother youthful, slim, and cheerful; his brother a successful broker; and his sister chic and batting away a swarm of ardent suitors like houseflies. He made decisions and altered outcomes, fortunately for the better. In so doing, he created a new universe, identical to his own, up until the time he changed the original succession of events.

But what about the idea for which Einstein is most famous? In September 1905, Einstein extended his work on special relativity and found that energy and mass share a relationship similar to that of space and time. This work culminated in a single revolutionary equation: $E = mc^2$. Translated into English, this means that matter and energy can be changed into each other. Conservation of energy still applies, so if a body emits a certain amount of energy, then the mass of that body must decrease by the same amount. Squaring the speed of light produces an enormous number. So a system with very little mass, like an atom, can release a tremendous amount of energy.

We may not have Doc's magical DeLorean at our disposal, but we do have Einstein and special relativity. By that token, we are already time travelers, because each of us is always moving through both space and time. And perhaps somewhere, in a parallel universe, the laws of physics as we know them don't apply, enabling scientists to conquer the obstacles to time travel, in either direction. Our parallel self is merrily hopping back and forth among the centuries, unlocking the secrets of both the future and the past.

22

Rocket
Man

One of the earliest science-fiction novels was Jules Verne's *From the Earth to the Moon*. Published in 1865, but set a few years after the resolution of the American Civil War, it centers on the disaffected members of the Baltimore Gun Club. The men are bored and restless: there are no new artillery weapons to construct and test; there is nothing left to blow to smithereens. They need a new purpose for their lives. So their leader, the veteran artilleryman Impey Barbicane, proposes that they build a gigantic cannon to fire a projectile to the moon. Barbicane's archrival, Captain Nicholl, wagers several thousand dollars that it can't be done. One would think his chances of winning were highly favorable, but ultimately the plucky Gun Club succeeds in launching not just a projectile but three men into space. Against all odds, Nicholl loses the bet.

Verne's notion of launching men to the moon with a cannon is implausible at best, but it does make for a ripping good yarn. And the underlying scientific principle is sound. According to Newton's third law of motion, for every action there is an equal and opposite reaction. Firing a bullet from a gun, or a projectile from a cannon, produces a powerful recoil force in the opposite direction, capable of propelling an object a considerable distance. That's the principle behind jet propulsion. Like airplanes, rockets are subject to the aerodynamic forces of weight (gravity), thrust, drag, and lift. Before an object can become airborne, it must counter inertia by generating sufficient thrust to overcome the pull of gravity. Once airborne, the object is subject to aerodynamic drag, which increases proportionally with acceleration, so greater and greater amounts of thrust are needed to keep it in the air. Eventually the drag will overpower the thrust, and the object will fall back to earth.

Rockets predate Verne by several thousand years. They are first mentioned in historical records in China in the late third century BC. During religious festivals, bamboo tubes were filled with saltpeter, sulfur, and charcoal (the basic components of gunpowder), then sealed and tossed into ceremonial fires, in hopes that the noise of the explosions would frighten away evil spirits. It was just a matter of time before someone noticed that when a tube was imperfectly sealed, it would skitter out of the fire, instead of simply exploding. When the tubes were properly sealed, the energy from the chemical reaction inside the tube had nowhere to go, and simply built up until it was sufficient to overcome the binding force of the bamboo tube—that is, it exploded. But a break in the seal allowed energy to escape, creating a recoil force that propelled the tube in the opposite direction.

By AD 1025 the use of rockets was an integral aspect of Chinese military tactics. At the beginning of the thirteenth century, when the Sung dynasty was being threatened by Mongol hordes, Chinese weaponry experts designed several types of projectiles, including explosive grenades and cannon, as well as rocket fire-arrows loaded with flammable material and iron shrapnel. Contemporary accounts describe how when a rocket was lit, it thundered so loudly that the sound could be heard for 15 miles. And the point of impact caused devastation for 2000 feet in all directions.

Rocket technology didn't reach Europe until about 1241, when Mongols used rockets against Magyar forces and succeeded in capturing the city of Buda (in what is now Hungary). The Mongols used rockets again in 1258 to capture the city of Baghdad. Within 10 years, the Arabs had caught on, using rockets against the French in 1268, during the seventh crusade. After that, the technology spread like wildfire: an army would use rockets against an opponent, the opponent would in turn use them against another enemy, and so on. By 1300 the Germans and Italians had also added rockets to their arsenals; the

Italians even adapted rockets back to their initial use, as fireworks. The French, Dutch, and English soon followed suit, and by the eighteenth century rocketry was commonplace as a military weapon—and occasionally as an entertainment spectacle—across Western Europe.

Rockets also spread to the new world. During the American Civil War, the Union army made a less than successful attempt to use rockets against Confederate forces defending Richmond and Yorktown, Virginia, in 1862. Once ignited, the rockets skittered wildly across the ground and through the legs of several tethered mules. One rocket detonated under one of the mules, lifting it several feet off the ground. Though otherwise unharmed, the mule promptly deserted the Confederate army.

Early propellants were weak, so the rockets were mostly used as sloppy bombardment weapons and were often less effective than a cannonball. Rockets and cannons used the same underlying principle of propulsion, but rocket trajectories were much more difficult to control. In a solid-fuel rocket (the type sold in model rocket kits), the fuel and oxidizer are mixed together into a solid propellant and packed into a cylinder casing; it burns only when ignited. The main advantage is the ease of storage. But solid fuel doesn't burn fast enough to provide the rockets with enough power to overcome drag and get them sufficiently airborne. And while solid-fuel rockets can accelerate more quickly at liftoff, they can't be throttled in flight; they simply burn until all the fuel is gone. So their trajectories can't be controlled once the launch is complete. This made them unsuitable for sending a rocket into space.

Scientists intent on traveling to other worlds had to turn their attention to developing better liquid fuels. In a liquid-fuel rocket, the fuel and oxidizer are stored separately and pumped into a combustion chamber, where they mix and burn. Therefore, liquid fuels require complicated piping and pumping equipment, adding significantly to the weight. The trade-off is that they have greater propulsive thrust, although it takes time to build up after ignition. Liquid-fuel rockets can also throttle their power, providing better control during flight.

Verne's novel, with its lovably quirky characters and plethora of technical details, inspired dreams of space travel in countless boys, some of whom grew up to be pioneering rocket scientists. One was the

Romanian-born Hermann Oberth, who in 1923 wrote a prophetic book, *The Rocket into Interplanetary Space*. As a boy, Oberth had read Verne, and by age 14 he had already envisioned how a recoil rocket could propel itself through space by expelling exhaust gases from a liquid fuel from its base. He realized that the higher the ratio between propellant and rocket mass, the faster his rocket would be. But as the rocket expended fuel, its mass would remain the same while the engine's ability to provide sufficient thrust would weaken. He thought that using stages of rockets might be a solution; each section would be jettisoned as its fuel was depleted, since it would then be just dead weight. "If there is a small rocket on top of a big one, and if the big one is jettisoned and the small one is then ignited, then their speeds are added," he wrote.

First, someone had to invent a practicable liquid-fuel rocket. On October 19, 1899, 17-year-old Robert H. Goddard was staying with his family at the suburban home of friends near his hometown of Worcester, Massachusetts, when he climbed into an old cherry tree to prune its dead branches. Instead of doing so, he gazed at the sky and daydreamed about the possibility of launching a rocket capable of traveling to Mars. "I was a different boy when I came down from that tree," Goddard—who became a professor of physics—later said. He marked the day as a private anniversary for the rest of his life, to remember the time when he dedicated himself to the realization of space flight. Like many boys before and after him, Goddard was fascinated by fireworks and other pyrotechnics, which he believed held the key to realizing his dream of going to Mars—a dream that was intensified when he read Verne's novel, along with H. G. Wells's *The War of the Worlds*.

Goddard first attracted public notice for his work in rocketry as a student at Worcester Polytechnic Institute in 1907, when he produced a dramatic cloud of smoke from a powder rocket fired in the basement of the physics building. Somehow he escaped expulsion. Instead, school officials, impressed by his ingenuity, granted him several leaves of absence to further study rocket propulsion.

As early as 1909, Goddard conceived of a liquid-fuel rocket, and he continued his research after joining the faculty at Clark University in Worcester. In 1919 he published a paper suggesting that a demonstration rocket should be flown to the moon. At the time, such a notion seemed preposterous, and Goddard found himself dismissed as a crank by the general public, despite the scientific merit of his paper. After a number of attempts at designing a rocket, Goddard chose gasoline as the fuel and liquid oxygen (lox) as the oxidizer. Below –297 degrees Fahrenheit, oxygen is a liquid at atmospheric pressures. At higher temperatures, it vaporizes, and produces tremendous pressure in closed containers. Goddard used the pressure of this gas to push both liquids simultaneously from their tanks, through separate pipes, to the combustion chamber, where they mixed and burned.

It took 17 years, but Goddard finally achieved flight of a liquid-fuel rocket on March 16,1926, at his aunt Effie's farm in Auburn, Massachusetts. The entire contraption was roughly 12 feet high. Only a few steps were necessary in the countdown and launch. First, an assistant used a blowtorch on a long pole to reach up and ignite the black powder. He then closed the vent on the lox tank—thereby raising the tank's pressure—and quickly lit the alcohol-soaked cotton in the burner. Next, Goddard piped oxygen from the cylinder to the pressurized propellant tanks, forcing the gasoline and lox to the combustion chamber, where the igniter was still burning. The rocket motor fired. When the motor's thrust exceeded its weight, it rose a few inches from the ground, tethered only by the hose. With a long rope, Goddard pulled a hinged rod that yanked the hose away, and the rocket was free to fly.

Goddard recorded the occasion in his diary: "It looked almost magical as it rose, without any appreciably greater noise or flame, as if it said, 'I've been here long enough; I think I'll be going somewhere else, if you don't mind.'"—After 2½ seconds of flight, the fuel was expended and the rocket fell to earth 184 feet away. It had reached an estimated speed of 60 miles per hour, and a height of 41 feet. Although the rocket was rudimentary and a far from practical design, it validated Goddard's basic concept.

Around the same time in Germany, there was another boy who dreamed of space flight and grew up to be a rocket scientist: Wernher von Braun. Like Goddard's, his imagination was fired by reading—in his case, Oberth's visionary book. Von Braun was just as enthralled by explosives and fireworks as Goddard, to the chagrin of his father, who considered his son a juvenile delinquent. As a teenager, von Braun strapped six skyrockets to a red toy wagon and set them off. The wagon traveled five blocks, streaming flames, before the rockets exploded, destroying the wagon, and von Braun was arrested. Despite this inauspicious beginning, he went on to earn a PhD in physics in the late 1930s. Within two years he found himself heading Nazi Germany's military rocket development program, working alongside his hero, Oberth. Von Braun invented the V-2 ballistic missile, first launched on October 3, 1942. It followed its programmed trajectory perfectly and landed on target 120 miles away. This would be the ancestor of practically every missile used today.

The Germans launched these ballistic missiles against London in 1944, but the breakthrough came too late. By April 1945, the German army was in full retreat and Hitler had committed suicide in his bunker in Berlin. Von Braun and his entire team of rocket experts had been ordered executed to prevent their capture, but von Braun's brother smuggled them to the safety of nearby American forces before the SS could carry out the order. They then went to work for the Allied forces. Von Braun had never wanted to build weapons; he dreamed of using rockets to travel into space, and once the war officially ended, he finally had his chance.

Goddard died in 1945, and thus missed the dawn of the modern space age. Von Braun emigrated to America. He went on to head NASA's Marshall Space Flight Center and develop the Saturn V class of "super-booster" rockets, based on Oberth's concept, which would finally land a man on the moon. On July 16, 1969, amid unprecedented media coverage, the Apollo 11 mission was launched from the Kennedy Space Center in Florida. Commander Neil A. Armstrong, Command Module Pilot Michael Collins, and Lunar Module Pilot Edwin "Buzz" Aldrin found themselves inside the most powerful rocket ever built, on a three-day journey to the moon. Each morning mission control would wake the crew and give them a news report, and color TV telecasts were made daily as the astronauts went about their routine chores.

The spacecraft reached the moon on July 19, and the next day Armstrong and Aldrin boarded the lunar module—the Eagle—and headed down to the surface, announcing their arrival to Mission Control: "Houston, the Eagle has landed." At 10:39 PM EST, Armstrong opened the outside hatch of the lunar module and slowly made his way down the 10-foot ladder. Fifteen minutes later, the entire world was watching as Armstrong placed his left foot on lunar soil and declared, "That's one small step for man; one giant leap for mankind."

Apollo 11 was followed by five more lunar landings and, in 1973, the launch of Skylab, the world's first space station, but none of them proved as spectacularly successful as that historic mission. By the 1980s, the space shuttle had become the jewel in NASA's crown and remained so well into the 1990s, although the program has been somewhat tarnished by the disasters that destroyed the *Challenger* and the *Columbia.*

Today, NASA is looking beyond the moon. In 1997, the unmanned Pathfinder spacecraft reached the surface of Mars, and beamed back to Earth the first images of the Martian surface. NASA followed that mission with two robotic rovers, Spirit and Opportunity, launched in June and July 2003; as of August 2004, they were still probing the surface of Mars, taking geological samples and helping scientists create a detailed topological map of the planet. It's a shame Goddard did not live to see them.

23

That Darn Cat

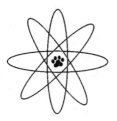

1935

Schrödinger's quantum cat

A rriving in Hollywood for a physics conference, the quantum physicist Ruth Baringer wants to check into her hotel room at the Rialto in time for the welcoming reception. But she seems to have stumbled into some kind of alternative reality where the ordinary rules don't apply, and something as simple as getting a hotel room becomes as randomized as a roll of dice. This is the bizarre world of Connie Willis's award-winning science-fiction short story, "At the Rialto."

Tiffany, the receptionist, is a model-slash-actress with the brain of a pea-slash-amoeba, who is working at the hotel only to pay for her organic breathing lessons, and never seems to have a reservation for anyone. Even those lucky enough to acquire a room key may find their room already occupied, or nonexistent. Order a doughnut at the local diner, and you might get the sea urchin pâté instead, nestled on a bed of radicchio. Lectures are never in their scheduled locations at the appointed times—you're as likely to find a new age spirit guide as a physicist on the podium. And no one can find the keynote speaker, Dr. Sidney Gedanken, who is rumored to have received a lucrative funding grant to develop a new paradigm for quantum mechanics.

The name of Willis's keynote speaker is a physics in-joke. *Gedanken* is German for "thought." And one of the most famous thought experiments in history was the brainchild of the Austrian physicist

Erwin Schrödinger: the paradox of Schrödinger's cat. His fanciful notion has piqued the interest of philosophers and appalled cat lovers ever since it was first proposed in 1935 as an illustration of the inherent absurdity of the quantum realm. It has also captured the imagination of countless artists, musicians, and writers. The novelist Robert Anton Wilson is best known for his *Schrödinger's Cat Trilogy.* Choreographer Karole Armitage created a ballet called *The Schrödinger Cat.* In the 1980s the pop group Tears for Fears had a song by that title. A rock band in upstate New York and an urban hip-hop a cappella group in Austin, Texas, both call themselves Schrödinger's Cat.

While its implications are profound and far-reaching, Schrödinger's thought experiment is quite simple. Place an imaginary cat inside a box along with a vial of cyanide, and close the box. The vial is connected to a Geiger counter near a uranium atom, a highly unstable element. This radioactive atom has a 50 per cent probability of decaying and emitting an electron. If that occurs, it will trigger the Geiger counter, the vial of cyanide will be broken, and the cat will be instantly killed. If not, then no cyanide will be released, and the cat will live. To find out what happened to the cat, we have to open the box. But what is the state of the cat before we do so? The beleaguered feline exists in two superimposed potential states: both dead and alive at the same time. The act of observation is what ultimately decides its fate. Special relativity also relies to some extent on human perception: an event cannot be said to have taken place until light reaches the observer, confirming that something has happened. But the outcome is pre-determined; a relative observer has no power to change or affect what has already happened. In contrast, according to quantum mechanics, observation actually determines what has happened.

This kind of paradoxical absurdity makes the quantum world unique. In her novel *Gut Symmetries,* Jeanette Winterson memorably describes Schrödinger's cat as "the new physics belch at the politely seated dinner table of common sense." At the dawn of the twentieth century, when scientists realized that atoms were not the smallest form of matter, they uncovered an entirely new realm of the universe that was governed by very different physical laws. Albert Einstein envisioned the fabric of space-time as smooth and curved, an orderly grid neatly laid out in predictable geometric patterns. But according to quantum theory, at the atomic level, the fabric of the universe is lumpy and chaotic, and subatomic particles are always in motion. The situation is similar to looking at a piece of paper: it appears to be smooth, but place it under a microscope and suddenly you can see the rough texture. Calculating how all those tiny particles interact is not possible using either classical or relativistic equations. At that size scale, gravity becomes irrelevant and electromagnetism is insufficient.

New York City's geography provides a simple illustration. For the most part, Manhattan has a neat grid of streets, much like Einstein's clean, well-ordered space-time: numbered avenues run north-south, and numbered streets run east–west. But the order begins to break down below Fourteenth Street. There are tiny side streets with no apparent connection to the sensible grid system. Seventh Avenue inexplicably becomes Varick Street, and there's an unlikely intersection between West Fourth and West Tenth Streets in Greenwich Village. Something similar happens when you start to move down to the level of individual atoms. The sensible, well-ordered rules of classical physics break down, and the bizarre world of quantum mechanics emerges.

Let's take a look at what would happen to the laws of physics if Manhattan's geography truly mirrored the macro- and atomic scales. Fire a cannon in Central Park, and you could easily calculate the projectile's trajectory and predict where it will land, using the equations of classical physics. Fire a cannon in Battery Park at the lower (quantum) tip of Manhattan, and you would have to consider all possible trajectories the projectile could take, then average them out to determine the most likely spot it would land. But it would still be a statistical approximation. There is no way to determine the exact landing site until the projectile actually lands.

So when Willis's Ruth tries to find the Rialto's breakfast buffet to get a cup of coffee and a doughnut—a seemingly straightforward endeavor—she has only a certain probability of succeeding, because she is caught in her own quantum paradox. According to quantum mechanics, the universe at the atomic level is neither orderly nor predictable; instead, it is a game of chance and probability. Einstein never fully accepted this approach, once famously objecting, "God does not play dice." But contrary to all expectations, it appears that at the atomic scale, God is the master crap-shooter in a microcosmic casino. Everything in the quantum world is ruled by chance. All scientists can do is calculate the odds and assign probability to any given experimental outcome.

One of the weirdest aspects of life on the quantum scale is the fact that all particles sometimes behave like waves. We have already seen in Chapter 20 how light can be both a particle and a wave, just as Tiffany can be both a model and an actress while moonlighting as a hotel receptionist. The phenomenon isn't limited to photons. Electrons, protons, and even neutrons all share a similar dual nature. The French physicist Louis de Broglie extended the notion of particle-wave duality to electrons in 1925. Even a simple water wave is granular at the atomic level, he reasoned, since it is composed of the coordinated motion of a horde of water molecules. All "particles" and all "waves" were in fact a mix of both. Because their "wavelengths" were so small, such "matter waves" wouldn't affect the macro world; their effects would appear only at the atomic scale.

De Broglie also suggested a possible experiment to detect the matter waves: shining a beam of electrons through a crystal. The atoms in a crystal are very well-ordered and arranged in a lattice-like structure, providing a built-in array of "slits" narrow enough to scatter the electron waves. When the experiment was performed in 1927, as de Broglie had predicted, the electrons didn't reflect from the surface along straight lines, like tiny balls. Instead, they acted like a water wave rippling outward when a stone is dropped into a pond. If the water wave encounters a barrier with two openings, the wave splits into two different parts as it passes through the barrier and makes striking spiral patterns as the waves recombine on the other side. The same kinds of "interference" patterns occur with a light wave passing through an array of small openings. An unobserved electron also acts like a wave when it is sent through two narrow slits, forming the telltale interference pattern. Any act of observation or measurement to determine which path the particle actually follows destroys that interference pattern. Essentially, observation forces the electron to "choose" one path or another, so that it acts like a particle traveling through one slit, instead of a wave breaking apart to go through both slits and recombining on the other side.

The British physicist John Gribbin likens the wave nature of electrons to the vibrations of a violin string. A taut string will vibrate in such a way that the two ends will remain fixed points while the middle wriggles back and forth. The frequencies of those vibrating sound waves will correspond to a specific musical note. Touch the center of the string, and each half will vibrate in the same way, with the center at rest. This corresponds to a higher, related note on the scale, called a harmonic. The violin string can vibrate only at specific frequencies corresponding to musical notes that harmonize with each other. This concept is similar to Niels Bohr's contention that electrons can circle atoms only in very specific quantum energy levels—much as light comes in neatly packaged photons with specific energies. In Bohr's model of the atom, the atomic nucleus is made of protons and neutrons, and electrons circle around the nucleus in a concentric array of fixed orbits with specific energies. Photons are emitted whenever an electron falls to a lower energy level.

Bohr's German protégé, Werner Heisenberg, didn't like his mentor's notion of well-defined electron orbitals around the atom. This was because in order to observe an electron to determine which orbital it inhabits, you have to shine light on it. That light must have a very short wavelength, and hence extremely high energy. The electron will absorb that extra energy and immediately jump to a higher orbital, leaving its original position. Locating an electron's precise position is thus impossible, since any act of observation or measurement will change that position. This is the basis for Heisenberg's uncertainty principle: the more accurately you measure one property, such as an electron's position, the less accurately you are able to measure another, such as its energy.

Schrödinger surmounted this problem in 1926, when he devised a set of equations that described the probability of where an electron could be. He reenvisioned Bohr's fixed orbitals as cloud-like regions of space around the atomic nucleus. The densest area of a given cloud has the greatest probability of finding the electron; the least dense area has the least likelihood. The physicist Sidney Perkowitz of Emory University likens the concept to determining how many times a specific card in a deck of playing cards will appear over many random drawings. You can say that the ace of hearts will be drawn in roughly $1/_{52}$ of the trials, but you can't predict when that will happen. Only the act of actually drawing a card will convert potential choice into a specific outcome.

Schrödinger's mathematical "wave function" describes all possible states of an electron, or of that darn quantum cat. In the same way that looking inside the box determines whether the cat is alive or dead, the act of measuring somehow focuses a matter wave—which carries all physical possibilities—into a particle with very specific properties. The electron has no precise location. Instead, it exists in a superposition of probable locations. Only when an experimenter observes the electron does the wave function "collapse" into one specific location.

Schrödinger was as upset as Einstein about the implications of quantum mechanics, once declaring, "I don't like it, and I'm sorry I ever had anything to do with it." His fictional ill-fated feline implies that nothing is real unless it is observed. The notion is very zen: if a tree falls in the forest, and there is nobody around to hear it fall, does it make a sound? In its strictest interpretation, quantum mechanics dictates that even the most objective of experimental outcomes is dependent upon human observation. Set up an experiment with a particle detector in place, and light will show up as a particle. Repeat the same experiment with a wave detector in place, and the light will exhibit wavelike behavior.

Particle-wave duality isn't the only bizarre aspect of life at the subatomic scale. Ruth's quantum conundrum at the conference is further complicated by the presence of David, a fellow physicist with whom she shared a romantic interlude the year before. Determined not to be distracted by love from the serious business of science, she throws up all sorts of barriers to keep herself from giving into David's irrepressible charms. Yet somehow he manages to bypass those obstacles.

Subatomic particles exhibit similar behavior in something called "quantum tunneling." In the 1950s, physicists devised a neat little system in which the flow of electrons would hit a barrier and stop: most electrons lacked sufficient energy to surmount the obstacle. But some electrons didn't follow the established rules of behavior: they simply tunneled right through the energy "wall." How could this be? Again, the electron dons its wavelike personality. We can imagine the electron as a wave of a certain height trying to surmount a tall barrier. Even if the wave is shorter than the barrier, quantum mechanics says that it has a small probability of seeping through the barrier and making it to the other side, where it can then reassume its particle identity.

When subatomic particles collide, they can remain invisibly connected, even though they may be physically separated. The state of Schrödinger's cat somehow "knows" the state of the radioactive atom, even at a distance; they are inextricably interlinked. So if you measure the state of one, you will know the state of the other without having to make a second measurement. This is called "entanglement." Once Ruth and David have "collided," they become entangled. "It didn't matter where the electron went after the collision," Ruth muses in resignation after David has tracked her down in a local diner. "The other electron would still come and rescue it from the radicchio and buy it a donut ... It was as if they were eternally linked by that one collision, even if they were on opposite sides of the universe."

This happens even though Einstein's theory of special relativity dictates that there is always a delay of at least the speed of light. Entanglement somehow finds a way to pass the cosmic speed limit. A scientist could entangle a pair of particles, separate them by vast distances, and then instantly deduce the state of one by measuring the state of the other. Einstein called it "spooky action at a distance." This is not just a theoretical construct: physicists have experimentally demonstrated quantum entanglement in the laboratory with both photons and clusters of atoms.

For example, in 1997 researchers at the University of Geneva in Switzerland entangled two photons and then sent them flying in opposite directions down fiber-optic lines to detectors nearly 7 miles apart. When the researchers measured the properties of one photon, it had an instantaneous effect on the other. Of course, only the particles instantly "know" their relationship; outside observers don't have the built-in advantage of entanglement. Scientists have to discern the precise state of two particles by measuring them at either end and then communicate the information to each other by telephone or e-mail. And even in this era of high-speed communication, that information cannot be transmitted faster than the speed of light.

None of these phenomena are apparent on the larger scale of our everyday lives. The charm of Willis's story lies in her ingenious "What if" scenario: What if quantum effects did become manifest at the macro scale? Hollywood replaces Schrödinger's cat as the new paradigm for the quantum world. Maybe, Ruth surmises, electrons are like

Tiffany the model-slash-actress, working part-time at being electrons to pay for their singlet-state lessons. Maybe the quantum realm could be seen in the Capitol Records building (shaped like a stack of records), or the seemingly random patterns of celebrity handprints and footprints preserved in cement along the star-studded walk in the courtyard of Grauman's Chinese Theater. Physicists are just beginning to understand this brave new subatomic world. As David says, "I think we know as much about quantum theory as we can figure out about May Robson from her footprints."

24

Copy
That

October 1938
First xerographic copy

The infamous Mormon forger-turned-murderer Mark Hoffman produced hundreds of masterfully faked historical documents. One of his masterpieces was a bit of priceless Americana: the so-called "Oath of a Freeman." It was the first page printed on the very first American printing press in 1639, and an authentic document would fetch more than $1 million in the rare books and manuscripts market. Arguably one of the most gifted forgers in history, Hoffman used all kinds of ingenious tricks and arcane chemistry to turn ordinary paper and ink into supposedly valuable memorabilia. But for his bogus "Oath," forged in 1984, he began with a simple Xerox machine.

Hoffman photocopied the relevant pages from a stolen library copy of the Bay Psalm Book, printed on the same press a few years later, which contained a reproduction of the oath. This enabled him to copy accurately not only the specific typeface used on that early press, but also the design for the decorative border. He used a razor blade to cut out the letters and designs he wanted, then painstakingly glued them to a piece of paper to create the oath. He photocopied this composite and took the copy to a local engraver, who used it to cut a zinc printing plate. A page could then be printed from the engraving. Hoffman's ersatz "Oath" was good enough to confound even document experts for the Library of Congress. Though suspicious, they could find no evidence to prove conclusively that the document wasn't genuine. The matter wasn't resolved until Hoffman confessed, after being arrested for murder.

As long as there have been methods for printing and copying images and text, there has been forgery; each is an art unto itself, but they are closely linked. For millennia, the primary means of reproducing texts was to copy them out painstakingly by hand. But errors inevitably crept in as the scribes became tired or bored. In AD 175 the Chinese emperor Ts'ai Yung feared that inexact copying of vital cultural texts would eventually lead to their being lost. So he wrote his own authorized version of six classic texts and had it carved into stone outside the gates of the state academy. Exact copies could be made by placing paper over the inscription and rubbing it with ink.

Between the fourth and seventh centuries of the Christian era, rubbings gave way to wood block printing, also invented in China, although it is believed to have derived from ancient Babylonian seals used to stamp impressions into wax or clay as authentication. The text was carved into a wooden printing plate, which was inked and used to print the text. The first mention of woodblock printing was an imperial decree of 593, in which the emperor Wen Ti ordered the printing of Buddhist images and scriptures carried as charms by believers. By the end of the ninth century, printed books were common all over China—not just Buddhist scriptures, but Confucian classics, dictionaries, and mathematical texts.

In the mid-eleventh century, a Chinese alchemist named Pi Sheng invented a form of movable type made from an amalgam of clay and glue. One Chinese character was carved in relief on each small block of moistened clay, which was then baked to make the type hard and durable. The pieces of type could be glued to an iron plate and coated with a mixture of resin, wax, and paper ash. The plate was then heated and cooled to set the type, which could be easily detached by reheating the plate. The invention never really caught in China, mostly because the complexity of the language required a very cumbersome machine: as many as 30,000 Chinese ideograms were needed to make a complete font.

Exactly when ancient printing methods first spread to western Europe is unknown, but historians believe that when the explorer Marco Polo visited China in the thirteenth century, he saw printed books and brought the knowledge back with him. For centuries, monks in European monasteries painstakingly copied out classical and biblical texts by hand onto pages of vellum, which were often lavishly illustrated. Books were viewed as consecrated objects, so the monks who devoted their lives to copying out chapters of the Bible believed they were making the word of God manifest in the world. (Ironically, this is the same reason Chinese printing methods never spread to the Muslim world. Muslims believed the handwritten word of the Quran to be sacred, and that reproducing it by any other means than hand-copying would be blasphemous.) Monasteries had very poor cataloging systems—hardly anyone knew where specific books were located—and the texts were considered far too valuable to be made available to scholars.

But the Western world was changing rapidly with the rise of universities and increasing literacy. To meet the growing demand for books, stationers' shops were established to provide paper, ink, and texts for copying by students. Whenever a student needed a text for a class, he could go to a stationer's shop and copy it by hand. But again, this led to a compounding of inaccuracies in texts. Finally Johannes Gutenberg conceived of his own scheme for movable type and adapted a common wine press to create the first European printing press, with the help of a local wood turner named Konrad Saspach. If a letter broke down, it could be replaced, and when the printing of copies of one page was finished, the type could be reset for the next page or book. His first production was the Gutenberg Bible, printed in 1455.

The printing press helped usher in the Renaissance in Western Europe. But by the early twentieth century, people still sought a cheap, efficient means of making copies of images or documents that didn't require such a labor-intensive process. An American physicist turned patent clerk named Chester Carlson would invent one. Carlson was an only child, born into abject poverty. By age 14 he was working after school and on weekends for a local printer to support the family, since his father—a barber by trade—could no longer work, having developed crippling arthritis in his hands. His mother succumbed to tuberculosis a few years later. Carlson worked his way through a nearby junior college, then earned a degree in physics from CalTech two years later. But the country was caught in the Great Depression, and jobs were scarce. Carlson applied to 82 companies before landing a job in the patent department of an electronics firm in New York.

Bored with the routine of the patent office, Carlson turned to inventing in his leisure hours. At the time, there were no quick, efficient means of mass reproduction, so it was difficult to make sufficient copies of patent documents to distribute in the office. There were only two options, both of which were costly and time-consuming: either send the patents out to be photographed, or laboriously type new ones. Carlson envisioned a device that would make copies of a document in seconds. He started by researching common copying techniques at the New York Public Library.

The hectograph had first appeared in the 1870s. It used a gelatin pad to absorb ink from an original, and blank sheets were then pressed against the pad to produce impressions, much as in the modern stamp pad. In 1887 Thomas Edison invented the mimeograph, in which a waxed stencil was fitted around an inked drum to transfer an image to paper when the drum was rotated. Carlson was particularly interested in the "spirit duplicator" that appeared in 1923, the precursor to the "ditto" machines that were still being used in some high schools as recently as 1980. A master sheet with a waxed back could be typed or written upon. The master was wrapped around a cylindrical drum. As the drum turned, the master was coated with a layer of duplicating fluid to slightly dissolve the dye on the master. The image could then be transferred to paper pressed against the drum. It was capable of making up to 500 copies before the print became too faint to read.

Yet all these were "wet" processes. Carlson sought a dry technique. He found the answer in photoconductivity. A few years earlier, the Hungarian physicist Paul Selenyi had discovered that certain materials, such as sulfur, conduct electricity in light but act as insulators in the dark. When a photoconductive layer is exposed to light, the energy of the photons liberates the electrons and allows current to flow through. Carlson combined this concept with the fact that materials with opposite electrical charges attract one another. He realized that if the image of an original document was projected onto a photoconductive surface, current would flow only in those areas that had been exposed to light. The areas where print would appear would be dark, producing a copy of the original. If he could find a way to get dry particles to stick to a charged plate in the pattern of the original image, he could make "dry reproduction" work.

He called his concept "electrophotography"; today we know it as xerography—from the Greek *xeros* ("dry") and *graphos* ("writing")—or more commonly as photocopying. After filing a preliminary patent application for his idea, Carlson began conducting experiments in the kitchen of his apartment in Queens. When his wife objected, he moved his experiments to a rented room in Astoria, and hired a young German physicist in need of work as his lab assistant, Otto Kornei.

The first xerographic copy was made on October 22, 1938. The two men prepared a sulfur coating on a zinc plate, and Kornei printed "10–22–38 Astoria" in India ink onto a glass microscopic slide. They pulled down a shade to darken the room. Then they transferred a charge to the sulfer-coated zinc plate and laid the glass slide over it. Next, Carlson and Kornei shone light onto the two pieces, just for a few seconds, to trigger the photoconductive effect. They removed the slide and dusted the plate's surface with lycopodium powder—the same substance used to lift fingerprints, and the precursor to modern-day toner. When the excess powder was blown off, the zinc plate was "inscribed" with an almost exact replica of the "10–22–38" notation from the original glass slide, etched out by the remaining particles of powder. Like any good scientist, Carlson repeated the experiment several times, just to make sure his success wasn't a fluke and that his process really worked. He even figured out how to make his copy permanent by making the transfer onto wax paper and then heating it to melt the wax.

Carlson took his invention around for several years, trying to find a company willing to make the capital investment required to develop

First "Xerox" copy

the process into a useful product. More than 20 companies passed on the opportunity—including IBM, Kodak, General Electric, and RCA. "How difficult it was to convince anyone that my tiny plates and rough image held the key to a tremendous new industry," Carlson later recalled. Finally, in 1944, Carlson caught a break. Battelle Memorial Institute agreed to partner with a small photo paper company to commercially develop a prototype based on Carlson's process. That small company was Haloid, which would later become Xerox Corporation.

The first photocopiers were introduced in 1949. They were not an immediate success. For one thing, they used flat plates instead of rotating drums, and required 14 separate steps and some 45 seconds to produce a single copy. It took another 11 years before the first auto-mated office copier appeared: the Xerox 914. The machine was so popular that by the end of 1961 Xerox had earned nearly $60 million in revenue, and by 1965 revenues had jumped to more than $500 million. Carlson ended his years a very rich man.

Today's photocopiers still work in very much the same way. The concept is similar to how a camera takes a picture, but instead of a negative image created on photosensitive film, a photocopier creates a real image from a pattern of positive charges left after exposure to light. Nor is there any need for chemical processing to develop film. The copier produces an image with dry ink (toner), heat, and regular paper. Inside is a rotating drum covered by a layer of photoconductive material such as selenium, germanium, or silicon. A lightbulb flashes light onto the original document and the pattern of the image is pro-jected onto the drum. The dark areas of the original document absorb light, and the corresponding areas on the rotating drum are not illu-minated. Charged particles remain only in places that were not exposed to light, and these positively charged dark areas attract the negatively charged toner. The modern laser printer is little more than an ordinary photocopier, except that a laser beam replaces the reflected light used in an ordinary copier.

One unforeseen consequence of Carlson's invention is that the advent of color photocopiers has greatly reduced the level of skill required to counterfeit currency; most counterfeiters are between ages 16 and 21. Always precocious, Mark Hoffman committed his first act of forgery at 14, subtly altering the face of a coin to increase its value in the collectors' market. Since 1993, Xerox has made photocopiers with built-in software that can tell when a document being copied is currency. One model automatically colors the entire page cyan blue when it detects a bill as an original, rendering the copy useless. Another causes a blue screen to come up on the user interface, warning that reproduction of currency is forbidden, and the copier automatically shuts down.

Another innovation is electronic paper. Corporate offices use huge amounts of paper; globally about 280 million tons are used annually. Electronic paper is essentially a reusable display, like Xerox's new SmartPaper. It's made of two sheets of thin plastic with millions of two-color beads surrounded by oil so that they can rotate easily. When voltage is applied, the beads rotate from black to white, as need be, to produce patterns on a page—much like pixels in a computer monitor. The technology is ideal for retail signage because it can be used over and over and updated wirelessly, and it may also provide a handy alternative to today's electronic books. Users may even be able to "write" on SmartPaper by waving a wand over it to download documents or e-mail.

Among the many inventive delights of the Harry Potter books are the collectible cards that come in packages of chocolate frogs, and have photographs of famous wizards whimsically disappearing and reappearing—not to mention the characters that inhabit paintings at Hogwarts and interact with students and teachers. Gryffindor's Quidditch captain Oliver Wood demonstrates game tactics using a special map marked with arrows that wriggle like caterpillars. Fred and George Weasley inherit the Marauder's Map that shows everyone's location in Hogwarts and can be wiped clean after every use. And there is Lord Voldemort's magical Hogwarts diary, which plays a critical role in the plot of *Harry Potter and the Chamber of Secrets*. The books are pure fiction, but according to the British physicist Roger Highsmith, author of *The Science of Harry Potter*, all these things could well be printed on some form of electronic paper in the future.

25

Life During Wartime

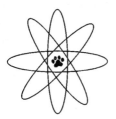

July 1945
The Trinity test

In Stanley Kubrick's satirical masterpiece, *Dr. Strangelove, or How I Learned to Stop Worrying and Love the Bomb* (1964), a general at a U.S. Air Force base goes mad and sets in motion a course of events that will lead to a man-made Armageddon. The paranoid general sees a communist conspiracy everywhere, even believing that the fluoridation of the water supply is a scheme to poison all Americans. He drinks only distilled water and pure wood grain alcohol—which might account for his madness. So he sends instructions to bomber pilots armed with nuclear warheads to attack their assigned targets in Russia, then locks down the base and cuts off all communication.

The general's intent is to force the United States to quit shillyshallying and commit itself to all-out war with the Soviet Union. "War is too important to be left to the politicians," he proclaims. "They have neither the time, the training, nor the inclination for strategic thought." Unfortunately, the Russians have a "doomsday device," set to trigger atomic bombs around the world automatically in the event of nuclear attack, and capable of annihilating all life on the planet. The president of the United States and his cabinet members—including Dr. Strangelove, a former Nazi scientist who has emigrated to America—scramble frantically to abort the mission, and they very nearly succeed. But one lone bomb is dropped, and it sets off the doomsday device. The result is a multitude of thermonuclear mushroom clouds.

"Inside the horror of Nagasaki and Hiroshima lies the beauty of Einstein's $E = mc^2$" the novelist Jeanette Winterson observes in *Gut Symmetries*. That equation is indeed the underlying principle behind thermonuclear weapons. Mass is simply a highly concentrated form of energy, and mass and energy can be changed into each other. Energy is conserved, so if a body emits a certain amount of energy, then the mass of that body must decrease by the same amount. If you think of energy and mass as two forms of currency, the speed of light squared would be the enormous exchange rate between them—akin to getting 448,364,160,000,000,000 euros to the dollar. At those rates, millions of American college students could tour Europe in style on just a few dollars. And a system with very little mass, such as an atom, can potentially release a tremendous amount of energy.

The protons and neutrons in an atom's nucleus are held together by something called the strong force. In nuclear fission, that force is overcome, and an atom splits apart, forming two or more atoms and releasing a neutron. But if you add the masses of the two new atoms and the neutron together, the sum is less than the mass of the original atom. Where does the missing mass go? It is released in the form of heat and radiation (kinetic energy). Something similar happens in nuclear fusion—the source of energy in the sun and other stars—which occurs when lightweight atomic nuclei find themselves in an extremely hot environment. The high temperature transfers a great deal of kinetic energy to the atomic nuclei, which then bounce around and collide with each other so furiously that they fuse together to form a heavier atom. The mass of the new atom is still less than the combined masses of the original atoms, and the missing mass is also released as kinetic energy.

Almost 30 years after Einstein thought up $E = mc^2$, the Parisian physicists Irène and Frédéric Joliot-Curie captured the conversion of energy into mass in a photograph. Around the same tune, physicists in England managed to break apart an atom. As Einstein had predicted, the fragments had slightly less mass than the original atom, and they emitted excess energy equivalent to the missing mass. However, the release of energy was too gradual to be of much practical use. The Hungarian physicist Leo Szilard proposed that bombarding atomic nuclei with extra neutrons would make the atoms chemically unstable and trigger a chain reaction to release energy much more quickly.

By late 1938, Nazi Germany had annexed Austria; it then invaded Czechoslovakia in March 1939 and Poland in September 1939. These invasions came on the heels of Japan's invasion of China two years before, so Britain and France had little choice but to declare war to counter the aggression. When the United States entered the war, the federal government ramped up efforts to develop an atomic bomb under the so-called Manhattan Project—located not in New York City, but in Los Alamos, New Mexico. Headed by the physicist J. Robert Oppenheimer, the Manhattan Project brought together some of the greatest scientific minds in the world to surmount the daunting technical challenges of harnessing the power of the sun. And Szilard's notion of a chain reaction became the basis for the first atomic bomb.

Not every element can undergo induced nuclear fission. The most promising candidate was a rare isotope, uranium (U) 235. If two atomic nuclei of the same element have the same number of protons but a different number of neutrons, they are isotopes. Some isotopes are stable; others are not, and will decay to form other elements. Uranium 235 has three fewer neutrons, making it less stable than standard uranium. If a free neutron collides with a U 235 nucleus, the nucleus will absorb it instantly. This makes the atom unstable, and it will split into two smaller atoms, with the excess mass converted into energy. A pound of pure U 235 is smaller than a baseball, but its potential power is equivalent to that of 1 million gallons of gasoline. The fission process also creates more neutrons, which in turn trigger other fission events. This unstable condition is called "supercriticality."

Uranium 235 seemed like the perfect fissile material. The only problem was that it was extraordinarily expensive and difficult to isolate from uranium 238, another isotope, which makes up the vast bulk of uranium. Early estimates on how much would be needed to achieve critical mass—an amount sufficiently large to support an explosive chain reaction, but small enough not to explode spontaneously—ran into tons. But scientists hadn't considered how much faster the reactions would occur if fast-moving neutrons were injected into pure U 235 nuclei. In that case, mere kilograms would be needed to achieve critical mass. So an atomic bomb was feasible.

To avoid triggering fission prematurely, the fuel components must be kept separate, much as in liquid fuel rockets, where the fuel and oxidizer are kept separate until the energy from the chemical reaction is needed. Also, the two subcritical masses must be brought together very quickly, before the energy released by the initial fission blows them apart and stops the chain reaction. The simplest way is to fire one mass into the other with a gun. "Little Boy," the bomb dropped on Hiroshima, was a gun-triggered fission bomb. "Fat Man" dropped on Nagasaki, was an implosion-triggered fission bomb in which the two subcritical masses are surrounded by explosives. When fired, the explosives create a shock wave that compresses the masses at the bomb's core, triggering the fission reaction. (The odd propensity for naming lethal bombs is satirized in *Dr. Strangelove:* the nuclear warheads are labeled "Hi There" and "Dear John.")

By mid-1945, it was time to test a prototype bomb. The scientists at Los Alamos chose a secluded spot in a central New Mexico desert, called Jornada del Muerto ("Walk of the Dead"). Nicknamed "Gadget," the first atomic bomb was hoisted to the top of a 100-foot tower. Just before sunrise on July 16, 1945, at 5:29:45 AM, Gadget was detonated.

The blast vaporized the steel tower, and within a few minutes the mushroom cloud rose to more than 38,000 feet. Several observers standing toward the back of a shelter were knocked flat by the shock wave, and the heat of the explosion melted the sandy soil around the tower to form a mildly radioactive glassy crust known as "trinitite." The shock wave broke windows as far as 120 miles away and was felt at least 160 miles away. A military policeman on hand for the test described the heat as being "like opening up an oven door, even at 10 miles." Another witness compared it to seeing two sunrises: "There was the heat of the sun on our faces. Then, only minutes later, the real sun rose, and you felt the same heat to the face."

You'd think a nuclear explosion would be difficult to hide, but the so-called "Trinity" test was not made public until August 6, just after Little Boy had been detonated over Hiroshima, killing some 70,000 to 130,000 people. Three days later, Fat Man was detonated over Nagasaki, killing 45,000 people. Japan surrendered on August 14. It was a huge victory for the Allied forces, but one gained at tremendous cost: Fat Man and Little Boy literally shattered worlds. A survivor of Hiroshima later recalled that the blinding flash turned everything "sheer white," and described a ring of light "like a halo around the moon" spreading outward, followed by a column of flame reaching to the sky that detonated "like a volcanic explosion in the air."

The image of a mushroom cloud billowing upward has become indelibly etched in our collective consciousness, and Kubrick is not the only filmmaker to find that image visually compelling. In James Cameron's film *Terminator 2: Judgment Day*, Sarah Connor is haunted by a recurring nightmare: as mothers and children cavort in a playground in Los Angeles, there is a sudden flash of light that radiates outward, vaporizing everything in its path. This is a reasonably accurate depiction of the effects of a nuclear blast. The explosion creates a high-pressure shock wave of gamma rays that radiates intense heat. Just how much damage occurs depends on the distance from the center of the bomb blast. Anything near the center is vaporized by the heat of the explosion. As it travels outward, the shock wave will collapse buildings and other structures, resulting in flying debris. Then there is radioactive fallout: clouds of fine radioactive particles of dust and bomb debris that rise up with the mushroom cloud and are scattered to the winds. If a nuclear bomb were dropped on Thirty-Fourth Street in Manhattan, the shock wave would radiate out over most of the island, destroying nearly everything in its path.

One survivor of Nagasaki returned home to find the carbonized bodies of his parents, and no trace of his brother and sister. Within a week, his gums bled black and he was running a high fever from the radiation exposure. He was lucky compared with some. Like X rays, gamma rays are a form of ionizing radiation that can damage or kill living cells. Studies of survivors of Hiroshima and Nagasaki revealed such symptoms as nausea and severe vomiting; cataracts in the eyes; severe hair loss; and loss of blood cells. There was also an increased risk of later developing leukemia or cancer, and of infertility or birth defects.

It wasn't just atoms that were split with the detonation of the first atomic bombs; the physics community experienced its own kind of fallout. After Trinity, Hans Bethe famously observed, "The physicists have known sin. And this is a knowledge which they cannot lose." He meant that physicists had so relished the challenge of building a bomb that they forgot the implications of what they were creating. Immediately after the successful test, Richard Feynman recalled finding his colleague Robert Wilson sitting despondently while others were wildly celebrating. "It's a terrible thing that we made," Wilson told Feynman.

At least one physicist on the Manhattan Project remained untroubled by any prickings of conscience: Edward Teller. Unlike Bethe, Wilson, and Oppenheimer, who were appalled at what physics had wrought, Teller refused to accept the perceived contradiction between the results of science and issues of morality. Even before the Trinity test, he was intrigued by the feasibility—first suggested by Enrico Fermi—of using an atomic bomb to heat a mass of deuterium sufficiently to achieve fusion and ignite a thermonuclear reaction. It would be a "superbomb," roughly 700 times more powerful than Little Boy.

Teller wanted to pursue both options at Los Alamos, but his colleagues decided that building the simpler fission device was challenge enough, and the fusion project was abandoned. Teller's disappointment was palpable: he clashed with several of his fellow scientists, and even sulkily refused to perform a detailed calculation of Gadget's implosion, despite the fact that theorists like himself were in short supply. It wasn't until the Russians detonated their own atomic bomb that President Harry Truman ordered Los Alamos to proceed with developing a "superbomb." The first hydrogen bomb was successfully detonated on November 1, 1952, on Eniwetok Atoll in the Pacific Ocean. Oppenheimer, Fermi, and many other veterans of the Manhattan Project were vehemently opposed to the plan, and the result was a deep and bitter rift between two factions of atomic scientists.

Nowhere was this rift more apparent than in the infamous security hearings of 1954 to determine whether Oppenheimer was guilty of treason. Oppenheimer had several communist acquaintances dating back to the 1930s, and during an inquiry in 1942, under pressure, he had even implicated some of his friends as Soviet agents—testimony he later admitted was a "tissue of lies." His outspoken opposition to the hydrogen bomb did little to allay suspicion, and the Atomic Energy Commission (AEC) began compiling a file on his alleged questionable activities. This coincided with the onset of the McCarthy era, with its paranoid emphasis on rooting out "subversives." As chair of the Senate Investigations Subcommittee, Senator Joseph McCarthy unveiled a new policy under which not only did a government employee have to be judged "loyal," but his or her background had to be "clearly consistent with the interests of national security."

During the hearings, Teller testified against Oppenheimer, telling the commission, "I would prefer to see the vital interests of this country in hands that I understand better and therefore trust more." Many scientists felt that this was an unforgivable betrayal of a colleague, and ostracized Teller. The AEC found Oppenheimer innocent of treason, but ruled that he should not have access to military secrets. His security clearance was revoked on the grounds of "fundamental defects of character," and for communist associations "far beyond the tolerable limits of prudence and self-restraint" expected of those holding high government positions.

The lone dissenting opinion among the members of the AEC came from Commissioner Henry DeWolf Smyth, who found no evidence that Oppenheimer had ever divulged secret information during nearly 11 years of constant surveillance. Smyth, a professor of physics at Princeton University, believed that the charges against Oppenheimer were supplemented by "enthusiastic amateur help from powerful personal enemies," and concluded that, far from being a communist subversive, Oppenheimer was "an able, imaginative human being with normal human weaknesses and failings." Einstein and 25 colleagues at Princeton joined the Federation of American Scientists in protesting the AEC's decision. But the damage had been done.

"I am still asked on occasion whether I am not sorry for having invented such a terrible thing as the hydrogen bomb. The answer is, I am not," Teller wrote in the May 22, 1998, issue of *Science* magazine. He pointed to the fact that several decades later, America won the cold war, and claimed (with characteristic hubris) that his work had "played a significant role" in the outcome. The fact that detonating thermonuclear weapons led to the arms race in the first place seems to have escaped his notice.

Kubrick has said that the sheer insanity of the concept of nuclear "deterrence" was the reason he made *Dr. Strangelove* a comedy. But is it really all that laughable? In the film, the Russians develop their doomsday device as a "deterrent" to forestall a nuclear attack, so naturally the Americans also want one, even in the face of imminent annihilation. "We cannot have a doomsday gap!" the chief of staff declares. The diabolical ingenuity of the device lies in its simplicity: all it takes is one madman with the will to use it.

26

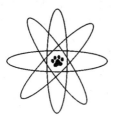

Gimme
Shelter

June 2–4, 1947
The Shelter Island conference

On the afternoon of June 1, 1947, commuters in New York City were startled by the sight of a long line of policemen on motorcycles, sirens blaring, escorting a bus to Shelter Island, located between the twin forks of Long Island. The passengers on the bus weren't movie stars, star athletes, or powerful politicians, but a group of 24 theoretical physicists, en route to a small scientific conference at the Ram's Head Inn. What had they done to warrant such preferential treatment? At the time, physicists were national heroes, since they had helped win World War II by building the first atomic bombs. A member of the local chamber of commerce, John C. White, had arranged for the escort as a gesture of respect and gratitude. He had served in the Pacific during the war, and credited the physicists on the Manhattan Project with saving his life.

Once the physicists arrived, they went straight to work. "They roam through the corridors, mumbling mathematical equations, and eat their meals amid the fury of technical discussions," a bemused reporter for the *New York Herald Tribune* wrote on June 2. To a town more accustomed to hosting corporate conferences, they seemed like a superior alien race. Rumors buzzed among the local residents that the physicists were working on another bomb, but the purpose of the Shelter Island conference was less tangible than that. They were there to explore nothing less than the future of theoretical physics, which had languished during the war.

Many considered the field to be in crisis. The noted physicist I. I. Rabi declared that except for the bomb, the 18 years since the birth of quantum mechanics had been "the most sterile of the century," in terms of theory. His colleague Murray Gell-Mann later recalled that, at the time, "theoreticians were in disgrace." In physics, the predictions of any successful theory must agree with experimental results. Quantum mechanics explained the strange behavior of electrons in matter at the atomic level, which in turn explained all of chemistry and the way different substances behaved. But for some reason, it didn't quite work when physicists tried to apply those same equations to explain the interactions between light and particles of matter, such as electrons. The predictions frequently came out wrong. And try as they might, theoreticians had thus far failed to rectify the problem.

That might seem like nitpicking to those outside the field, but physicists take their theory very seriously. It is, after all, the underpinning of everything they do. So they organized the Shelter Island conference. Attendance was by invitation only, and the participants included the finest scientific minds of the era, many of them veterans of the Manhattan Project. For all the collective brainpower, the conference resulted in no immediate advances, no new major insights—and certainly no new kind of thermonuclear bomb. Yet it is deemed by science historians to be a turning point for postwar theoretical physics, because it laid the groundwork for a successful quantum theory of light. Although it goes by the cumbersome name of quantum electrodynamics (QED), that theory is the crown jewel of modern physics.

If you have never heard of QED, you're not alone. Not very many people outside physics have a clear idea of what it is or why it's important, and even some physicists admit that they know almost nothing about the specifics. QED isn't necessarily taught until the third or fourth year of graduate study in physics, when students are finally able to follow the basic equations. Yet the theory is so significant that three men—Richard Feynman, Julian Schwinger, and Sin-Itiro (Shinichiro) Tomonaga—won the Nobel Prize in physics in 1965 for remedying its one major flaw.

So how come practically everybody on the planet can cite $E = mc^2$, yet hardly anyone knows about QED? For starters, QED can't be summed up in a simple formula. Unlike the elegant simplicity of

Albert Einstein's equation, the equations of QED are incredibly convoluted. In the 1930s, when physicists first tried to apply quantum mechanics to the theory of light, the algebra to calculate just a single quantity often spilled over seven or eight lines. It's also much harder for a layperson to grasp why physicists felt a quantum theory of light was necessary. Classical physics treats light as a wave; quantum mechanics treats it as particles that travel like waves. Isaac Newton's laws of motion describe the way objects behave under the influence of gravity, but Einstein's general theory of relativity explains what gravity is: it arises from the warping of space-time by the presence of massive objects. James Clerk Maxwell's electromagnetic equations do something similar for light, but they treat light as a wave. Physicists still needed a quantum theory that could explain the behavior of light as a particle, and that's what QED does. Ever since the ancient Greeks, people could describe how light behaves. QED explains why it behaves that way, down to the level of individual photons and electrons.

Although he was one of three physicists who were honored with the Nobel Prize for their work in QED, Feynman became its poster child because he was so good at explaining the abstract concepts to the general public. He always told his audiences that the difficulty of QED was no reason to be afraid of it. "My physics students don't understand it either," he assured them. "That's because I don't understand it. Nobody does." It never bothered Feynman if he didn't understand something. "Not knowing," he was fond of saying, "is much more interesting than believing an answer which might be wrong." Feynman, who grew up in Far Rockaway, Queens, loved to solve puzzles from a very young age. When he was in high school, a student would often come to him in the morning with a geometry problem, and Feynman would work on it until he figured it out. Over the course of the day, other students would present him with the same problem, and he'd solve it for them immediately, earning a reputation as a "super-genius." His entire career in physics was about solving ever more complicated puzzles.

Feynman was nothing if not original. He joined the Manhattan Project in the early 1940s, while still a graduate student, evincing a mischievous penchant for breaching the Los Alamos security systems just for fun. He taught himself the art of safecracking and picked the locks on vaults containing the most sensitive secrets involved in building an atomic bomb. He never took anything. He just left taunting notes behind bemoaning the project's lax security. In his later years, he developed a passion for painting and playing bongo drums, frolicked in hot springs with nudists, and frequently visited strip clubs, often scribbling equations on paper napkins between performances. He claimed that the atmosphere cleared his thinking. Many women found his lusty joie de vivre irresistible, and he racked up an impressive number of romantic conquests in the intervals between his three marriages.

This was hardly behavior becoming to a "serious" physicist, but Feynman believed that nothing was worth doing unless it was also fun. It was a lesson he learned after falling into a creative slump while a young physics professor at Cornell University in the late 1940s. There were very good reasons for his intellectual lethargy. His beloved first wife had died of tuberculosis a few days before the Trinity test; he was still reeling from the loss. And while the Manhattan Project had proved to be a great success for physicists in terms of ending a horrible war, it also ushered in an era of suspicion and fear, dividing the physics community as never before. Add in the pressure from public expectations to top such a spectacular achievement, and it's little wonder Feynman lacked his usual innovative drive.

One day a depressed Feynman was in the university cafeteria when some prankster threw a plate into the air. The plate was decorated with Cornell's signature red medallion seal on the rim, and as the plate went up, Feynman saw it wobble. He also saw the red medallion spinning around. The plate's wobble was clearly faster than its spin, and Feynman set about calculating the exact ratio between the two. He found that when the plate is angled very slightly, its wobble is twice as fast as its spin. Asked by a colleague about the point of the exercise, he exclaimed, "It has no importance—it's just for fun!" His depression lifted, and Feynman found himself once more enthusiastically tackling the problem of QED, with that same sense of play. "It was like uncorking a bottle," he later recalled.

What was the problem? Early attempts at a quantum theory of light had turned out to be less than perfect. If you made an approximate calculation, you would get a reasonable answer. But if you tried to calculate more precisely, you would get an obviously wrong answer. Imagine punching numbers into a calculator. Most of the time, $2 + 2 = 4$. But every now and then, for no apparent reason, the calculator mysteriously gives an answer of 100,000, or even a long line of numbers stretching into infinity. This is known in mathematical parlance as a "singularity," and it's taboo in any theory that claims to describe real-world phenomena. Infinities exist only in the abstract realm of mathematics.

We saw in Chapter 23 that the laws of physics are much different at the quantum scale, where everything is ruled by chance and probability, and common sense is booted out the subatomic window. It's not enough to simply say that a ball rolled down the hill along a certain path. There are many different paths the ball could have taken to reach the same outcome. A quantum physicist must ask: Out of all the possible paths the ball could have taken, why did it "choose" that particular path? A simple process involving just one or two subatomic particles won't give an answer of infinity, because the number of possible paths is limited. But when you're dealing with a multitude of particles, there are many different ways for them to interact, each with the exact same outcome. QED's equations must consider every single option, and that's when the infinities pop up.

Here's another example. If a ray of light strikes the surface of a mirror, the angle at which it hits the surface is the same as the angle at which it leaves the surface. So the light must be traveling from its source along whatever path takes the least time to get to your eye—and common sense tells us that must be a straight line. The notion is not so much wrong as half right. In Peter Parnell's play *QED*, a fictionalized Feynman tosses a ball out to the audience, and says that if he knew how long it took the ball to travel, he could easily calculate the path it took. That's simple Newtonian physics. But if the ball were the size of an electron, or a photon of light, Feynman's calculations would have to take into account all possible paths it could conceivably travel. This is what really happens when light reflects off a mirror: every single photon is taking every possible path on its way to your eye. Nature combines all those possible paths into one average path, and that is the path we "see."

The blackboard in Feynman's office at CalTech—where he taught from 1950 until his death—had a phrase written at the top: "What I cannot create, I do not understand." Feynman couldn't understand quantum mechanics as it was presented in textbooks of the time. So he reinvented it. He streamlined the convoluted equations of QED by creating his own unique system: a series of simple line drawings known as "Feynman diagrams" that precisely illustrated how electrons interact with photons. Feynman envisioned those interactions as a theatrical play. The "actors" are the photons and electrons, and there are three basic actions from which all dramatic conflicts—all phenomena related to light and electrons—arise: (1) a photon bounces from one point to another; (2) an electron bounces from one point to another; and (3) an electron emits or absorbs a photon. Surfaces really don't have much effect on light at all. Incoming photons are absorbed by the electrons in the atoms inside a substance, and new photons are emitted. Those new photons are the ones we "see."

Feynman diagrams serve as mathematical shorthand, detailing all the possible ways the particles can interact to give rise to electromagnetic phenomena. In the simplest interaction, a pair of electrons move toward each other. The electron on the right emits a photon and is pushed outward, much as in the recoil that happens when you shoot a rifle. The emitted photon is then absorbed by the electron on the left, which is also pushed outward, just like what happens when a fired bullet hits a tin can, knocking it backward. It's all one big, elaborate

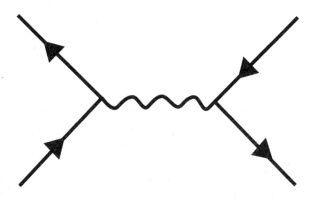

**Feynman diagram of a simple
photon–electron interaction**

subatomic game of catch. "Photons do nothing but go from one electron to another," Feynman said. "Reflection and transmission [of light] are really the result of an electron picking up a photon, 'scratching its head,' so to speak, and emitting a new photon."

This is why strange phenomena like partial reflection occur. If you're sitting inside with a lamp on in the room, and you look out the window in the daytime, you will be able to see not just the things outside through the glass, but also a dim reflection of the lamp. The light hitting the glass is partially reflected. Why does this happen? A piece of glass, Feynman said, "is a terrible monster of complexity—huge numbers of electrons are jiggling about." An incoming photon doesn't just bounce off electrons on the surface; it penetrates the entire glass, either bouncing off, or being absorbed by, the electrons it encounters along the way. But the result is always the same, even if the photon goes through several layers of glass. If light hits the surface of a piece of glass at a 90-degree angle, for every 100 photons that hit the glass, 96 go through (transmission), while 4 hit the glass and bounce back (reflection). Subatomic particles have no consciousness (as far as we know), so how does the photon weigh the myriad of options and "decide" which way to go? Scientists just don't know. Physicists can calculate only probability: out of every 100 photons, 4 will be reflected back. They can't predict which specific photons these will be.

Not everyone was a fan of the Feynman diagrams. Schwinger, who shared the Nobel Prize with Feynman, avoided using the diagrams altogether, sniffing dismissively that the stick figures had brought "computation to the masses." But being able to visualize the many possible interactions between electrons and photons has rescued many a physics graduate student from the brink of despair at ever being able to make sense of it all.

How did Feynman and his colleagues resolve the problem of infinities? It turned out that physicists were using the wrong yardstick. Electrons orbit atoms cloaked in a cloud of photons. The original QED equations based the values for the measured electron properties of mass and charge on the "bare" particle, not taking into account the surrounding cloud. But laboratory experiments can measure only the cloaked electron; scientists haven't found a way to experimentally get inside the cloud, so to speak, to measure the particle's "bare" properties. It's a bit like the difference in a person's weight naked versus fully clothed; the results don't match up well with the doctor's weight chart unless they are measuring the same thing. Feynman, Schwinger, and Tomonaga each independently discovered that by simply redefining the reference frame to reflect a "clothed" electron, they could get rid of the infinities.

Today, QED is the most rigorously tested theory in the history of physics, and the most successful in terms of the accuracy of its predictions. One physicist at Cornell has spent the last 30 years using it to calculate the electron's properties. The calculations ran for thousands of pages and ultimately required parallel supercomputers to complete. But the predictions matched experimental results to an accuracy of one part per billion. Feynman likened the feat to measuring the distance from Los Angeles to New York City so accurately that any remaining discrepancy would equal the thickness of a human hair.

Feynman developed a rare form of cancer in his sixties, and although successive surgeries bought him some time, by February 1988 his condition had worsened to the point where survival statistics were not in his favor. "It's like the hundred photons hitting the glass," Parnell's fictional Feynman muses in *QED* while considering his treatment options. "Am I one of the 96 that go through, or one of the four to bounce back? There's no way to know." In the end, he opted to let the disease take its course, but his last words were vintage Feynman. He emerged briefly from a coma to complain, "I'd hate to die twice. It's so boring."

27

Tiny Bubbles

1948

Reddi-Wip appears on the market

In 1948, a clothing salesman turned entrepreneur, Aaron ("Bunny") Lapin, came up with the brilliant notion of packaging real whipped cream in handy aerosol cans. First sold by milkmen in St. Louis on their morning rounds, Reddi-Wip was an instant success, becoming a symbol of postwar America's obsession with convenience, and earning Lapin the title "whipped cream king." Today, half of all the canned whipped topping sold in the United States is Reddi-Wip—about 4 million pounds are consumed annually in November and December alone—and in 1998, *Time* magazine listed this product as one of the twentieth century's top 100 consumer inventions, along with the pop-top can and Spam.

More than a great American capitalist success story, Reddi-Wip is a prime example of the ongoing human love affair with foam. Foams are a unique combination of solids, liquids, and gases, displaying properties of each depending on the surrounding conditions, and they exist in the universe at all size scales, from the quantum gravitational bubbles in the fabric of space-time, to the galactic structure of the cosmos itself. Foams permeate our daily lives, popping up in our favorite foods and beverages (bread and beer are the earliest examples, dating back at least to ancient Egypt), our personal grooming products, our household soaps, our great works of art, and even ocean waves as they crash upon the shore.

Foams are examples of so-called soft matter: they don't flow freely like a true liquid, but neither do they assume the definite shape of a crystalline solid like diamond. Foams arise when some form of mechanical agitation—say, a chef beating egg whites with a wire whisk to produce a fluffy meringue—thrusts air into a liquid, forming bubbles of many different sizes, although most are merely small fractions of an inch across. Initially, each bubble is a sphere: a volume of air encased in a very thin liquid skin that isolates the bubble from its neighbors. The bubbles owe their geometry to the phenomenon of surface tension, a force that arises from molecular attraction. The greater the surface area, the more energy is required to maintain a given shape. That is why the bubbles seek to assume the shape with the least surface area: a sphere.

The pull of gravity gradually drains the liquid downward, causing the bubbles to press more tightly against each other. As the amount of liquid in the foam decreases, the "walls" of the bubbles become very thin, so that smaller bubbles gradually are absorbed by larger ones. Over time, the tiny bubbles that make up foam become larger. The combination of these two effects is called "coarsening." As the coarsening continues, the bubbles begin to resemble soccer balls (a geometric shape known as a polyhedron), although they vary in size, shape, and orientation, as shown in the following figure.

Foams come in many different consistencies. Solid foams include packing peanuts and the polyurethane foam fillings in chair cushions, as well as naturally occurring foams like pumice. Whipped cream,

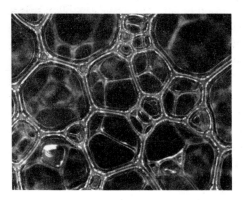

Bubbles in a foam

hairstyling mousse, and shaving cream are all examples of liquid foam, the result of air beaten into a liquid formula that contains, among other ingredients, a surfactant (an active surface agent): a collection of complex molecules that link together to stiffen the resulting froth into a substantial foam. The surfactant—usually fats or proteins in edible foams, or chemical additives in shaving cream or styling mousse—keeps surface tension from collapsing the bubbles by strengthening the thin liquid film walls that separate them.

Aerosol cans were invented in the 1940s as a delivery system for insecticides, but it didn't take long for entrepreneurs like Lapin to realize the enormous potential of the technology for consumer goods. The ingenuity of the basic mechanism lies in its simplicity: gas is mixed in with the liquid formula and packaged under pressure in the aerosol can. When the valve is opened, the mixture is propelled from the can by nitrous oxide (laughing gas), and the gas expands rapidly to create a foam. In Reddi-Wip, cream, which is 20 per cent fat, acts as the surfactant. In nondairy varieties, it is replaced by vegetable oil, which has an even higher fat content than cream, along with a frightening array of barely pronounceable synthetic additives: polysorbate 60, sorbiton monostearates, sodium steroil 2, lactylite, xanthan gum, and lecithin.

Lapin's company was also one of the first to produce shaving cream in a can, although he ultimately decided not to market that product. Gillette introduced its first canned shaving foam in the 1950s, using the same delivery mechanism as in Reddi-Wip. In fact, shaving cream and some brands of nondairy whipped toppings sometimes share ingredients, such as polysorbate 60 and xanthan gum. "You might want to avoid nondairy varieties in the future. Or consider shaving with the stuff instead," jokes Sidney Perkowitz, a physics professor at Emory University who is the author of *Universal Foam*.

Shaving foam took on another new twist in the 1970s with the introduction of shaving gels with delayed foaming action, like the best-selling brand Edge, manufactured by the S. C. Johnson Company. The patented gel formula is supersaturated and, of course, packaged under pressure. While they might not be visible to the naked eye, there are bubbles in the gel, and once the pressure is released, the bubbles begin to grow by diffusion, speeded up by mechanical agitation in the form of a hand moving the gel along the face or leg. The delayed foaming action is achieved by having less propellant in the formula than is used in other shaving creams, so that instead of an immediate explosion of mountainous foam, the process occurs more gradually.

The concept behind the supersaturated gels in Edge is also the basis for the handheld portable neutron detectors invented in the 1980s by Robert Apfel, a mechanical engineer at Yale University. The material used in Apfel's device is insensitive to protons and hence ideal for measuring neutrons. The foaming action is triggered only in the presence of ionizing radiation, which causes the bubbles to expand with an audible pop, changing the volume of the container's contents. The user receives instant feedback, determining the dose of radiation from a volume indicator; the higher the dosage, the greater the increase in volume. Since neutrons are unavoidable by-products of the high-energy X rays emitted by many medical accelerators, the detectors are especially useful in hospital settings.

No one is sure why we love foam so much, but human beings clearly derive intense pleasure from its unique frothy texture, finding all manner of inventive new uses for it. Perkowitz, who was inspired to write his book when he became fascinated by the complex structure of the frothy milk that topped his morning cappuccino, has toyed with the idea of one day producing a cookbook devoted entirely to foamy foods. In the mid-1990s, dance clubs across the nation began throwing "foam frenzy" parties, filling the dance floor with mounds of foam as much as 7 feet high. Revelers compared the experience to dancing in champagne. And a Web search for Reddi-Wip produces numerous references to its use in decidedly adult nonfood applications. Even Buffy, TV's Vampire Slayer, has remarked on the pleasures of foam while reflecting on her first collegiate drinking binge: "It's nice. Foamy. Comforting. It's ... beer."

That sense of comfort is one reason for Reddi-Wip's appeal, perhaps involving a Pavlovian response to childhood memories of family holiday gatherings, with copious amounts of Reddi-Wip spritzed onto pumpkin pies. "Everyone knows the sound of Reddi-Wip coming out of the can; it's something we all grew up with," says Mary Stansu, who conducts market research for the company. The aerosol delivery system also seems to appeal to consumers' sense of play. Jim Crotty, director of product development for Gillette's personal care division, says that its customers definitely prefer rich, thick lathers in shaving cream, rather than foams that are thin or runny, mentally associating the richness with all the desirable attributes of a pleasurable shaving experience.

Food chemists are very much concerned with what they call "mouth feel," and foam excels in that quality. The underlying premise is that as foam fills one's mouth, it contacts all the taste buds at once; other foods don't. There is even some scientific evidence suggesting that our taste buds are affected not just by the chemistry that makes up a flavor, but also by the amount of pressure exerted on them.

Yet foam isn't just a frivolity or convenience; it's also cutting-edge science. While the basic origin and macrostructure of foams are fairly well understood, our knowledge of their microstructure, mechanical properties, and how they flow—all of which are intricately connected—has been limited until quite recently by the inherent difficulties of studying such an ephemeral material in the laboratory. Foams are extremely fragile, with a short lifetime because of the coarsening process, and must be confined to a glass container of some sort in order to be studied—but this confinement itself affects the behavior of the foam. The walls of any container confining the bubbles create forces that press against those bubbles, changing their shape—and the shape of its bubbles determines a foam's properties.

Undaunted, scientists have devised all manner of ingenious methods to gather hard experimental data on foam. Leonardo da Vinci studied the foams that formed on the surface of flowing water, recording his observations on surface tension in his notebooks in 1508. The nineteenth-century Belgian physicist Joseph Plateau developed his own mix of soap, water, and glycerin to produce foams that lasted as long as 18 hours, suspending them on wire frames. The American botanist Edwin Matzke manually constructed his own foams in the laboratory, bubble by bubble, putting them into beakers and examining their cellular structure under a microscope. While this might seem a bit obsessive-compulsive to those of us lacking the patience of Job, Matzke wasn't just killing time; he was intent on learning more about how cells pack together in living organisms, since plant cells have similar shapes. Andrew Kraynik at Sandia National Laboratory, who does computer simulations of the cellular structure of foams to predict their properties, says that Matzke's work was so thorough and so precise that his observations made in 1946 match remarkably well with Kraynik's own simulations of today.

Scientists know a great deal about the individual bubbles in a foam and how these bubbles "talk" to one another through simple friction. But when many bubbles clump together, the resulting foamy material exhibits many unexpected properties and behaviors. Liquid foams, for example, are roughly 95 per cent gas and 5 per cent liquid, yet they tend to be far more rigid than their components. This is due to a phenomenon called jamming. As a foam coarsens, there is less and less liquid to separate the individual bubbles, so the bubbles must press together more closely to fill the spaces left by the liquid. Because the bubbles are so tightly packed, they can't hop around each other. The more the bubbles are jammed together, the greater the pressure inside them grows—and, consequently, the more they take on the characteristics of solids.

Then there's the matter of shape. We've already seen that as a foam coarsens, its bubbles change shape from spherical to polyhedral because of the force generated when they press more closely together. The minimizing principle still applies: the bubbles seek out the most efficient way to pack themselves together, and this affects their shape. But unlike the perfect symmetrical structure of the hexagonal cells in a honeycomb, for example, the polyhedral bubbles in a foam are far from uniform: they can have anywhere from three to nine edges.

Mathematically, this is a very difficult shape to describe. In the nineteenth century, William Thompson (Lord Kelvin), who calculated absolute zero, came up with a shape that was eventually called Lord Kelvin's cell—a complex figure with six square faces and eight hexagonal faces that resembles a demented soccer ball. Kelvin built a wire model of this shape, which he called a tetrakaidecahedron (from the Greek for "14 faces"). It was a bold attempt, but Matzke, for instance, never found a single bubble adhering to Kelvin's proposed shape in all the thousands he studied. In 1994 Dennis Weaire and Robert Phelan of Trinity College in Dublin used a computer program to find a foam with even more efficiently packed bubbles. Their structure, which comprises two different shapes of equal volumes, is an improvement on Kelvin's model, but whether this really is the most economical structure possible has not yet been proved.

The biggest challenge facing the scientists of suds is creating predictive models of foam rheology: how, over time, it deforms and flows. For Glynn Holt, a researcher at Boston University, it's simply a matter of defying gravity. He uses sound waves to float individual drops of foam in midair, altogether eliminating the need for a glass container. With acoustic levitation, he can manipulate the suspended drop by altering the acoustic field, changing the drop's position and even squeezing it to cause the bubbles in the drop to oscillate. This is possible because sound waves travel throughout the foam, not just along the surface. A video camera hooked up to a monitor records the complex vibrational behavior of the foamy drop, which Holt then analyzes to glean specifics on the dynamic mechanical properties— most notably the point at which it begins to act more like a solid.

The physicist Douglas Durian took a somewhat different approach in 1991. He put some Gillette shaving cream fresh from the can into a small glass cell and shone a laser beam through one end of the cell, measuring the amount of light that emerged through the opposite side—a technique known as diffusing wave spectroscopy—to monitor the behavior of the foam over time. He found that the light's intensity fluctuated as the bubbles both consolidated and rapidly shifted position. As the foam shifted, its internal stresses grew, until groups of tightly packed bubbles suddenly snapped from one configuration to another, like a slow-motion avalanche. If he applied enough pressure, the bubbles would always keep rearranging themselves, flowing like a liquid while still maintaining their foamy nature.

Since foam is hardly a new phenomenon, why is it still the focus of so much research? One reason is that although physicists have become very adept at explaining hard matter such as crystals—the entire semiconductor industry is based on them—soft matter tells us a lot more about nature and biology. Unlike high-energy physicists or nuclear physicists, who believe that any grand unified theory will rest on a more complete understanding of the behavior of elementary particles, solid-state physicists are of the opinion that the key to unlocking the secrets of the universe lies in deciphering the structure of plant cells, or how physical laws emerge and evolve in biological systems— subjects as mysteriously complex as foam itself.

28

Mimicking Mother Nature

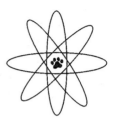

May 13, 1958
Velcro trademark registered

The sardonic talk show host David Letterman amused millions of viewers across the country when he flung himself through the air wearing a Velcro suit, to find out if he would stick to a corresponding makeshift "wall" of Velcro in his studio at NBC. (He did.) The stunt was inspired by a quirky barroom pastime of tossing Velcro-clad midgets at similarly constructed walls. This bizarre form of entertainment was obviously the brainchild of someone with little regard for political correctness and way too much free time, but it is a testament of sorts to America's enduring love affair with Velcro. Since its invention 50-odd years ago, Velcro has captured public imagination as a symbol of practical ingenuity—something so simple, and yet so useful, that many people wish they'd thought of it first.

Velcro belongs to a class of materials called polymers: molecules strung together into long chains. Usually made out of nylon, Velcro is used in sneakers, backpacks, wallets, jackets, watchbands, blood pressure cuffs, and toys like child-safe dartboards. It even helped to hold a human heart together during the first artificial heart transplant. The "stickiness" comes from its structure: examine the two strips of a Velcro fastener under a microscope, and you will see that one strip contains microscopic loops, while the other has tiny hooks that catch on the loops to fasten securely.

But Velcro is more than just a convenient fastener. It's also one of the best-known examples of biomimicry, a field in which scientists, engineers, and even architects study models and concepts found in nature, and try to use them to design new technologies. Janine Benyus, author of *Biomimicry: Invention Inspired by Nature*, defines it as a design principle that seeks sustainable solutions to human problems by emulating nature's time-tested patterns and strategies. "The core idea is that nature, imaginative by necessity, has already solved many of the problems we are grappling with," she writes. "Nature fits form to function, rewards cooperation, and banks on diversity."

The Eastgate Building in Harare, Zimbabwe, is a case in point. It is the country's largest commercial and shopping complex, and yet it uses less than 10 per cent of the energy consumed by a conventional building of its size, because it has no central air conditioning and only a minimal heating system. The architect, Mick Pearce, based his design on the cooling and heating principles used in the region's termite mounds, which serve as fungus farms for the termites. Fungus is their

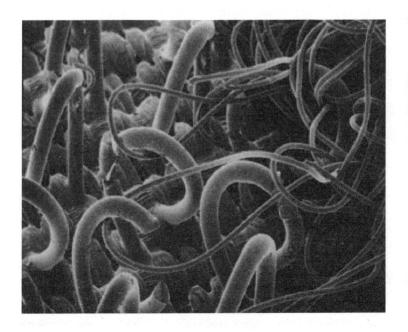

Velcro under a microscope

primary food source. While fungus is hardly on a par with delicate hot-house orchids, conditions still have to be just right in order for it to flourish. So the termites must maintain a constant temperature of 87 degrees F, in an environment where the outdoor temperatures range from 35 degrees F at night to 104 degrees F during the day. They do this by constructing a series of heating and cooling vents throughout their mounds, which can be opened and closed as needed over the course of the day to keep the temperature inside constant. The Eastgate Building relies on a similar system of well-placed vents and solar panels.

The history of human invention is filled with examples of bio-mimicry. Take humans' fascination with robots; so-called automata are first mentioned in ancient Greek texts. A Greek inventor named Ctesi-bius produced the first organ and water clocks with moving figures, and Hero of Alexandria detailed several automata used in theatrical perfor-mances. The Roman architect and engineer Vitruvius referred to automata in his writings, and developed the so-called "canon of propor-tions"—the basis for classical anatomical aesthetics. Leonardo da Vinci followed those principles when he designed what may have been the first humanoid robot in Western civilization: an armored knight that could sit up, wave its arms, and (because of its flexible neck) move its head.

Like most of his inventions, Leonardo's robot was never built. But by the eighteenth century, inventors were building a veritable menagerie of artificial mechanical animals, whose behavior simulated that of crea-tures in the natural world: peacocks, insects, dogs, swans, frogs, crayfish, ducks, even elephants. This was due in large part to the development of the art of watchmaking, combined with the prevailing notion of the era that the human body was largely a collection of similar intricate mech-anisms. Human-shaped android "musicians" were built, in consultation with doctors and surgeons to ensure that the various artificial organs were sufficiently lifelike. And a German inventor, Baron Wolfgang von Kempelen, built a "speaking machine" in which a bellows forced air through an artificial voice box to simulate human speech. The nobility found these creations highly diverting, but they also served to advance scientific understanding of the inner workings of human anatomy.

Both Leonardo and the Wright brothers studied the flight of birds when designing their flying machines, and Alexander Graham Bell used the principles of the human ear to design his telephone receiver.

Sonar was inspired by how whales, dolphins, and bats emit high-pitched sounds and analyze the returning echoes to help themselves navigate. Materials scientists study spider silk—which is 10 times stronger than steel would be at the same thickness—in hopes of one day imitating those properties in a man-made material. It would be ideal for parachute wires, suspension bridge cables, or weaving fabric for protective clothing. Or, who would have thought that slug mucus would have something to offer beyond its intrinsic "ick" factor? Yet it provides a useful model in the development of new synthetic lubricants capable of absorbing as much as 1500 times their weight in water; such lubricants could one day be used to combat friction in molecular-scale nanomachines. Researchers have even developed a kind of nanoscale Velcro, with hooks made of thin hollow tubes of carbon atoms, just a few atoms in circumference, called carbon nano-tubes. Nano-Velcro is 30 times stronger than the epoxy adhesives commonly used as fasteners in industry.

All this nature-based innovation would have delighted Velcro's creator, the Swiss engineer George de Mestral, who combined his love of invention with a passion for the great outdoors. Born in June 1907 in Switzerland, de Mestral received his first patent when he was 12, for his design of a toy airplane. His working-class parents couldn't afford to give him an advanced education, so he worked at odd jobs to pay for his studies at the École Polytechnic Fédérale de Lausanne, where he specialized in engineering. After he finished school, he took a job in the machine shop of a Swiss engineering company. In 1948, de Mestral took a two-week holiday from work to hunt game birds. While hiking with his Irish pointer in the Jura mountains, he was plagued by cockleburs, which clung relentlessly to both his clothing and his dog's fur. It was so difficult to disentangle the tenacious seed pods that de Mestral became intrigued by how they were constructed, and examined a few under a microscope. He noticed that the outside of each burr was covered with hundreds of tiny hooks that grabbed into loops of thread or, in the dog's case, fur. And it gave him an idea for a similar man-made fastener.

Most of the fabric and cloth experts he conferred with in Lyon, France—then the worldwide center for the weaving industry—were skeptical about the idea. But one weaver shared de Mestral's love of invention. Working by hand on a small loom, he managed to weave two cotton tapes that fastened just as strongly as the cockleburs had. De Mestral called the invention Velcro, from the French words *VELours* ("velvet") and *CROchet* ("hook"). The trademark name was officially registered on May 13, 1958. By then, de Mestral had quit his job with the engineering firm and obtained a $150,000 loan to perfect the concept and establish his own company to manufacture his new hook-and-loop fasteners.

Officially introduced in 1960, Velcro was not an immediate success, although NASA found it useful for getting astronauts into and out of bulky space suits. Eventually, manufacturers of children's clothing and sports apparel recognized the possibilities, and the company was soon selling more than 60 million yards of Velcro per year, making de Mestral a multimillionaire. He died in 1990, and was inducted into the National Inventors Hall of Fame nine years later.

Scientists continue to follow his example, looking to nature for innovative solutions to recalcitrant problems. Amid all the complicated equipment in the biophysicist Elias Greenbaum's lab at Oak Ridge National Laboratory in Tennessee is a common kitchen blender. That's because Greenbaum's cutting-edge research involves spinach. He goes through at least a pound of the stuff each week, and not in his lunchtime salad. Instead, he rinses it with water, stuffs it into the blender, and hits "puree." Once the spinach has been liquefied into a pulp, he uses a centrifuge to separate the various molecules, in much the same way that a salad spinner separates water droplets from lettuce leaves—but on a much smaller scale. It's an ingenious method he discovered for extracting tiny molecules in the spinach leaves that serve as minuscule chemical factories. These so-called "reaction centers" are the basic mechanism for photosynthesis, and a potential source of electrical energy.

Photosynthesis is the process by which plants convert carbon dioxide into oxygen and carbohydrates to make their own food. This involves a complicated series of chemical reactions, which are fueled by sunlight. First, the light energy converts water into an oxygen molecule, a positively charged hydrogen ion, and a free electron. Then all three combine to make a sugar molecule. Each of these reactions takes place in a reaction center. Greenbaum is interested in one particular reaction center—known as photosystem I (PSI)—which acts like a tiny photosensitive battery, converting light into energy by absorbing sunlight and emitting electrons. Spinach PSIs can generate a light-induced flow of electricity within a few trillionths of a second. It's not a lot of current, certainly not on a par with Popeye's impressive surge of strength after consuming a can of spinach. But Greenbaum thinks the amount of electricity is sufficient to one day run tiny molecular machines.

One of the first possible uses Greenbaum has found for PSIs is constructing artificial retinas for people who suffer from degenerative eye conditions. Such conditions usually result when the light-sensitive cells in the eye's retina are irreparably damaged. Greenbaum thinks he can replace the damaged cells with light-sensitive PSIs to restore vision. PSIs also behave like diodes, passing current in one direction but not the other. Diodes are the basis for constructing logic gates, essential building blocks of computer processors. The PSIs could serve the same purpose by connecting them into functioning circuits with molecule-size wires made out of carbon nanotubes. Spinach could even provide a new clean energy source. Greenbaum has discovered that he can get the PSIs to make pure hydrogen gas to power fuel cells, simply by sprinkling metallic platinum onto them. When light is shone onto the proteins, the metal catalyzes the splitting of water into hydrogen and oxygen, which can then react and give rise to heat, collected as energy. The only waste product would be water.

Iridescence is another natural phenomenon that scientists hope to exploit for new technologies. Anyone who has ever admired a peacock's brightly colored feathers might be surprised to discover that the colors don't come from pigments. Pigment molecules create colors by absorbing or reflecting certain frequencies of light, depending on the chemical composition of the molecules. But peacock feathers have only brown pigment (melanin). The bright colors we see arise from the inherent structure of the feathers, which have arrays of tiny holes neatly arranged into a hexagonal (lattice) pattern. This causes the light to refract off the surface in such a way as to produce the perception of color (be it blue, green, or red) in the human eye; which colors one sees depends upon the angle of reflection. Iridescence appears not only in peacocks, but also in beetles, dragonflies, the wings of kingfishers and certain species of butterfly, and the colorful spines of a marine worm called the sea mouse. In fact, any material with a periodic array of tiny

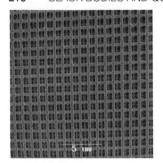

The precise lattice structure of a man-made photonic crystal. The scale marks 1 micrometre (micron): 1 millionth of a metre

holes or bumps may reflect some colors and absorb others—even if the material itself is colorless.

Scientists call such a structure a photonic crystal: an arrangement of atoms in a precise lattice pattern that repeats itself identically and at regular intervals. But nature doesn't produce crystalline structures with the level of precision that physicists deem necessary. They have to make their own version of these materials, atom by atom, to control and manipulate light. Light generally travels in a straight line, but if the atoms are organized precisely enough, they create what is known as a "photonic bandgap": certain frequencies of light can be blocked and reflected in new directions, and even turn a corner. The spacing of the atoms in the lattice structure determines which frequencies will be blocked. This is important for optical switching components in telecommunications, for example. Such structures could also lead to the development of novel clothing materials. A photonic crystal coat would absorb less heat from the sun, so the wearer would stay cooler in a hot desert. The coat could have a reversible pigmented side that absorbs heat to keep the wearer warm on chilly evenings.

The Victorian poet Gerard Manley Hopkins was an ordained minister who saw God's hand everywhere in nature and celebrated its diversity and beauty in his poetry. But it doesn't take a "believer" or a man of the cloth to appreciate the sentiment. Benyus, the author of *Biomimicry*, for one, would applaud Hopkins's impassioned declaration in one of his most famous sonnets: "Each mortal thing does one thing and the same/ . . . Crying *What I do is me; for that I came.*" Nature fits form to function, identity to purpose. That is the essence of biomimicry.

29

Energize Me

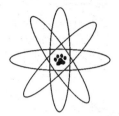

December 1958
Invention of the laser

In H. G. Wells's novel *The War of the Worlds*, residents of the quiet English countryside notice what appear to be meteors falling to Earth. The meteors turn out to be aliens, intent on conquering Earth and turning humans into slaves because their own planet, Mars, has become uninhabitable. At first they don't seem to be much of a threat: unaccustomed to Earth's stronger gravitational field, the Martians can barely move. But their advanced technology more than compensates for this effect. Soon 100-foot walking robotic tripods are laying waste to the countryside. Each has a heat beam that can melt or burn through almost any target it strikes. The British military, with nothing remotely similar in its arsenal, suffers a series of humiliating defeats. Things are looking rather grim for the human race, when nature intervenes. An entire race of superior alien beings is wiped out by the sniffles when the invaders catch a virus.

Wells's novel, first published in 1898, inspired a sizable sub-genre of science fiction centered on interplanetary warfare. In 1938, in the United States, Orson Welles adapted the story for a radio broadcast, changing the location of the alien invasion to the east coast. His drama was too convincing: many people in the audience, believing that Martians had indeed landed with the sole aim of eradicating the human race, were scared out of their wits. *The War of the Worlds* also prefigured the development of laser-based weaponry a good 50 years before the first working laser was invented.

Thanks in large part to the *Star Trek* and *Star Wars* series, laser-based weapons—from handheld lasers and light sabers to spacecraft shooting out intense beams of laser light to destroy enemy targets—are now a staple of popular culture. Laser-armed spacecraft are used by invading aliens in *Independence Day*. In the first Austin Powers movie, Powers's nemesis Dr. Evil awakens from a decades-long sleep and schemes to build his own gigantic laser weapon. His plan is to hold the world hostage for a whopping $1 million—until his lackeys point out that, inflation being what it is, perhaps $100 billion would be a more awe-inspiring sum. In the 1980s the Reagan administration, inspired by the thought of shooting incoming missiles out of the sky with laser beams, poured billions of dollars of federal funds into developing the so-called "Star Wars" strategic defense initiative—despite the fact that a panel comprising some of the best scientific minds in the world deemed existing technology woefully inadequate to make such a system a reality.

"Laser" is an acronym for light amplification by stimulated emission of radiation. It describes any device that creates and amplifies a narrow, focused beam of light whose photons are all traveling in the same direction, rather than being emitted every which way at once. Laser light contains only one specific color, or wavelength. Lasers can be configured to emit many colors in the spectrum, but each laser can emit only one. Because every photon is traveling in the same direction, the

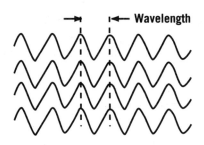

Coherent electromagnetic waves have identical frequency, and are aligned in phase.

A laser emits light waves with identical frequency that are perfectly aligned

light is tightly focused into a concentrated beam, unlike the light emitted from a flashlight, in which the atoms release their photons randomly in all directions. Anything that produces light—the heating element in a toaster; a gas lantern; an incandescent bulb—does so by raising the temperature of atoms. This causes the electrons orbiting the atomic nuclei to jump to higher energy levels; when they relax back into their ground state, the excess energy is released as photons. The wavelength of the emitted light depends on how many levels the electron had to drop.

There are many types of laser, but all of them have an empty cavity containing a lasing medium: either a crystal such as ruby or garnet, or a gas or liquid. There are two mirrors on either end of the cavity, one of which is half-silvered so that it will reflect some light and let some light through. (The light that passes through is the emitted laser light.) In a laser, the atoms or molecules of the lasing medium are "pumped" by applying intense flashes of light or electricity, so that more of them are at higher energy levels than at the ground state. Then a photon enters the laser cavity. If it strikes an excited atom, the atom drops back down to its ground state and emits a second photon of the same frequency, in the same direction as the bombarding photon. Each of these may in turn strike other energized atoms, prompting the release of still more photons in the same frequency and direction of travel. The end result is a sudden burst of so-called "coherent" light as all the atoms discharge in a rapid chain reaction. This process is called "stimulated emission."

Albert Einstein first broached the possibility of stimulated emission in a paper of 1917, but it wasn't until the 1940s and 1950s that physicists found a novel use for the concept. The physicist Charles Townes had worked on radar systems during World War II. Radar, invented in 1935 by the British physicist Sir Robert Watson-Watt, sends out radio signals at specific wavelengths, and when those signals strike a solid object, they are reflected back to the source. It's then possible to combine all the reflected signals to identify the object, as well as its position relative to the radar system. After the war ended, Townes turned his attention to molecular spectroscopy, a technique for studying the absorption of light by molecules; different types of molecules absorb different wavelengths of light. Just like radar, molecular spectroscopy bombards the surface of molecules with light and analyzes the ones that bounce back, to determine the molecule's structure. By

1953 Townes had patented a device he called a "maser," for microwave amplification by stimulated emission of radiation.

But the technique was limited by the wavelength of the light produced: in this case, the microwave regime of the electromagnetic spectrum. Townes noticed that as the wavelength of the microwaves shortened, the light interacted more strongly with the molecules; consequently, one could learn more about them. He thought it might be possible to gain even more information by developing a device that produced light at much shorter wavelengths, essentially extending the maser concept to the optical range of the electromagnetic spectrum. The best way to do this, he thought, would be to use molecules to generate the desired frequencies through stimulated emission.

Townes mentioned the idea to his colleague (later his brother-in-law) Arthur Schawlow, who proposed that the prototype laser be fitted with a pair of mirrors, one at each end of the lasing cavity. Photons of specific wavelengths would then reflect off the mirrors and travel back and forth through the lasing medium. By doing so, they would in turn cause other electrons to relax back into their ground states, emitting even more photons in the same wavelength. So only photons in the selected wavelength and frequency range would be amplified. The two men wrote a paper detailing their concept for what they called a laser, and published it in the December 1958 issue of a scientific journal, *Physical Review*, although they had yet to build a working prototype. They received a patent for their design two years later—the same year that the first working laser was built by Theodore Maiman at Hughes Aircraft Company.

Although Schawlow and Townes are the names most often associated with the laser's invention, numerous other people made vital contributions. Perhaps as a result, the issue of patent rights has proved to be fairly litigious. Gordon Gould, a scientist at Columbia University and later with Technical Research Group (TRG), sued for patent rights based on his research notebook, which had an entry dated and notarized in November 1957, describing his own design for a laser. Gould fought for decades, and in 1973 the U.S. Court of Customs and Patent Appeals ruled that the original patent awarded to Schawlow and Townes was too general and did not supply enough information to create certain key components of a working laser. Gould was finally granted patent rights.

Financially, there was a lot at stake: the laser's invention launched a multibillion-dollar industry. Lasers are used to remove unwanted tattoos or birthmarks, to correct vision defects, to cut through steel and other materials in industrial assembly lines, to scan prices in supermarkets and department stores, for optical communications and optical data storage, and in electronic devices such as CD and DVD players. None of these applications had even occurred to Schawlow and Townes, whose sole interest was in building a better instrument for scientific research.

There are several types of lasers. Schawlow and Townes designed solid-state lasers, which use crystals whose atoms are arranged in a solid matrix, such as ruby. CO_2 lasers emit energy in the far-infrared and microwave regions of the spectrum. This type produces intense heat and is capable of melting through objects. Dr. Evil coveted such a laser when he demanded "sharks with frickin' laser beams" on their heads to torture Austin Powers to death—only to be foiled because sharks are an endangered species. Imagine his disappointment if, in addition to having to make do with cranky mutated sea bass, he equipped them not with CO_2 lasers but with conventional diode (semiconductor) lasers. These are the type used in pocket laser pointers and CD and DVD players. They are not at all lethal.

CO_2 lasers are the most likely candidate for the fictional deadly "heat beams" used by the Martians in *The War of the Worlds*, or any other kind of laser weapons—with the exception of the light sabers in *Star Wars*. Unfortunately for aspiring Jedi knights, lasers would make very bad light sabers. For one thing, light will simply travel through space until it encounters an obstacle, so the weapon would have no fixed length. Thus it would inflict heat damage on virtually everything in the vicinity of its beam, not just a specific target. Second, two crossed laser beams won't clash and stop each other, so the masterful swordplay between Obi-Wan Kenobi and Darth Vader would be impossible. It's far more likely that each would sear the other in half on the very first beam pass, cutting a wide swath of destruction around them. The user would also have to avoid any kind of reflective surface, which would deflect the beam of light, possibly bouncing it back to fry off the user's face.

Lasers might have limitations as weapons, but they can still do some amazing things. For instance, lasers can be used to chill atoms down to unprecedented temperatures. In the mid-1980s, the physicist Steven Chu at Stanford demonstrated laser cooling by weaving a "web" out of infrared laser beams. He called it "optical molasses." The beams bombard target atoms with a steady stream of photons, whose wavelengths are carefully selected so that they will be absorbed only if they collide head-on with atoms. From the atoms' perspective, it's like running into the wind during a hailstorm. "No matter in what direction you run, the hail is always hitting you in the face," says Carl Wieman, a physicist at the University of Colorado's JILA facility (formerly the Joint Institute of Laboratory Astrophysics, but known simply as JILA since 1995). "So you stop." As the atoms slow down, they cool down to about 10 millionths of 1 degree above absolute zero.

In a laser beam, the light particles all have the same energy and vibrate together. Something similar happens to atoms in a form of matter called a Bose-Einstein condensate (BEC). BECs are named in honor of Einstein and the Indian physicist Satyendra Bose, who in the 1920s predicted the possibility that the wavelike nature of atoms might allow them to spread out and even overlap, if they are packed close enough together. At normal temperatures atoms behave a lot like billiard balls, bouncing off one another and off any containing walls. Lowering the temperature reduces their speed. If the temperature gets low enough (a few billionths of a degree above absolute zero) and the atoms are packed densely enough, the different matter waves will be able to "sense" one another and coordinate themselves as if they were one big "superatom."

With his colleague at JILA, Eric Cornell, Wieman embarked on a five-year quest to produce the first BEC, using a combination of laser and magnetic cooling equipment that he designed himself. While other groups were pouring research dollars into cutting-edge $150,000 lasers, he pioneered the use of simple $200 diode lasers. Using a laser trap, they cooled about 10 million rubidium gas atoms; the cooled atoms were then held in place by a magnetic field. This can be done because most atoms act like tiny magnets; they contain spinning charged particles (electrons). But the atoms still weren't cold enough to form a BEC, so the two men added a second step, evaporative cooling, in which magnetic fields forming a web conspire to kick out the hottest atoms so that the cooler atoms can move more closely together. This web works in much the same way that evaporative cooling occurs in your morning cup of coffee; the hotter atoms rise to the top of the magnetic trap and "jump out" as steam. Evaporative cooling was an old technique; the scientists at JILA simply tinkered with it until they got the low temperatures they needed.

Wieman and Cornell made history on June 5, 1995, at 10:54 AM, producing a BEC of about 2000 rubidium atoms that lasted 15 to 20 seconds. Shortly thereafter, Wolfgang Ketterle, a physicist at MIT, achieved a BEC in his laboratory. By September 2001, more than three dozen teams had replicated the experiment. In 2001, Wieman, Cornell, and Ketterle shared the Nobel Prize in Physics for their achievement.

The discovery launched an entirely new branch of physics. BECs enable scientists to study the strange, small world of quantum physics as if they were looking at it through a magnifying glass; a BEC "amplifies" atoms in the same way that lasers amplify photons. Among other things, scientists have used BECs to build an atom laser that drips individual atoms; someday, it could be used to etch tiny patterns on computer microchips. Other scientists hope to build atomic computer circuits that rely on the motion of atoms instead of electrons to store and process information. And in February 1999, researchers at Harvard University found that they could slow down light—which normally travels at 669,600,000 miles per hour—to just 38 miles per hour by shining a laser beam through a BEC. Two years later they briefly brought light to a complete stop.

Stopping light, even for a moment, raises the possibility of encoding information in atoms and transmitting it over light waves, in much the same way that we encode electrical signals onto radio waves. This is known as quantum communication, and it might eventually allow ultrasecure data transmissions—but only if we can stop a light wave long enough to change its state and send it on its way again. In *Star Trek*, every time Captain Kirk beams down to an alien planet, a beam of light rearranges distant atoms to match his body's molecular pattern back onboard the *Enterprise*. That, too, is an example of quantum communication. Achieving something similar with an entire human being across any distance is nowhere near a practical possibility, of course, but using BECs to stop light is a baby step in the right direction.

30

Small World

December 29, 1959
Feynman's classic Caltech lecture

Royalty has its privileges and its pitfalls. England's perennial monarch-in-waiting, Prince Charles, knows this firsthand, having been subjected almost since birth to intense scrutiny, prurient speculation, and often harsh criticism in every aspect of his life. On the other hand, his unique position gives him a public platform which allows him to call attention to issues that concern him—including perceived dangers linked to science and technology. In the 1990s he was concerned about genetically modified foods. In 2003, he turned his princely eye to the potential dangers of emergent nanotechnology. In particular, he fretted publicly that a voracious army of rogue self-replicating miniature robots would take over the planet and turn it into an uninhabitable wasteland by reducing everything in its path to an icky "gray goo."

If that apocalyptic scenario sounds a trifle familiar, that's because it's also the cornerstone of Michael Crichton's best-selling sci-fi novel *Prey* (2002). But the premise originated nearly 20 years earlier, when Eric Drexler, emeritus director of the Foresight Institute, published a book called *Engines of Creation*. Drexler envisioned a world transformed by microscopic nanoassemblers able to reproduce themselves and build anything, atom by atom, with unprecedented precision and no pollution. But he also described a downside: unless they were properly controlled, self-replicating nanomachines could run amok, growing to huge numbers so quickly that they would turn all life on earth to mush.

"Nanotechnology" is an umbrella term that describes any area of research dealing with objects measured in nanometers. A nanometer is a billionth of a meter and is analogous to the size scales for individual atoms and molecules. At that scale, quantum effects hold sway, and materials have chemical and physical properties different from those of the same materials in bulk. For example, carbon atoms can conduct electricity and are stronger than steel when woven into hollow microscopic threads.

The field is becoming big business, and everyone seems to want a piece of the action. Private corporations are pouring funds into nanotechnology-related research, on the basis of a prediction by the National Science Foundation that the field could grow into a $1 trillion industry in the United States by 2015. In a rare bipartisan gesture, President George W. Bush's budget for fiscal year 2004 allocated $849 million in federal funds to the National Nanotechnology Initiative founded by his predecessor, Bill Clinton. Congress passed a funding bill in 2003 allocating roughly $2.6 billion over the next three years for nanotechnology research. Japan is making comparable investments, and the European Union is not far behind.

Nanoparticles are already widely used in certain consumer products, such as suntan lotions, "age-defying" makeup, and self-cleaning windows that shed dirt when it rains. One company manufactures a nanocrystal wound dressing with built-in antibiotic and anti-inflammatory properties. On the horizon is toothpaste that coats, protects, and repairs enamel, as well as self-cleaning shoes that never need polishing. Nanoparticles are also used as additives in building materials to strengthen the walls of a structure, and to create tough, durable, yet lightweight fabrics. In 2004, scientists developed a superstrong flexible fiber that can conduct electricity, is as light as a cotton shirt, and is also bulletproof. This calls to mind the magical chain-mail undershirt which Frodo receives from the elves in J. R. R. Tolkien's *The Fellowship of the Ring* and which is capable of stopping a knife thrust. The metal, called mithril and once mined by the dwarves of Moria, looks like silver, is stronger than iron or steel, and is lighter than all three of them. It may well have been made of nanoparticles.

Today, nanotechnology is yielding some wonderful things. Michael Sailor, a materials scientist at the University of California, San Diego, has developed "smart dust": tiny particles of silicon sensors that can sniff out pollutants and toxins and signal a warning by visibly changing color. The "dust" can be painted onto walls or applied directly to protective clothing, and because the particles are so cheap to make, they can be used to form vast networks of sensors. Sailor alters the particles' chemistry to give them a porous, spongelike structure. The pores recognize and absorb molecules of certain toxic substances, even viruses and bacteria such as *E. coli,* and can change color when toxins are detected. For example, a particle of smart dust may have one green side and one red side. When the particle detects a drop of a toxin, the green side will attach to the drop and the red side will flash a warning.

In 1997, researchers at Cornell built the world's smallest guitar, about the size of a red blood cell and shaped like a Fender Stratocaster. Six years later, the same group created a slightly larger nanoguitar; this one was shaped like the Gibson Flying V model, and could actually be "played." Its "strings" are bars of silicon that vibrate in response to laser light. Pitches are determined by the length of the bars, rather than by the tension of the strings as in a real guitar. The nanoguitar can play only simple notes, but chords can be produced by vibrating more than one string at a time. The tones are far above any frequencies that humans can hear—17 octaves higher than those produced by a real guitar—but when the strings vibrate, they create interference patterns in the light reflected back, which can be detected and then electronically converted to audible notes.

The researchers had fun with this project, playfully posting some rudimentary "nanocompositions" on their website, including a short improvisational bit called "Cagey" because it is modeled on the work of the composer John Cage. But building nanoguitars isn't just an amusing laboratory exercise. The ability to make tiny objects vibrate at high frequencies opens the door to constructing smaller components for electronics systems. For example, cell phones and other wireless devices use the vibrations of a quartz crystal to generate a radio carrier wave for transmission or reception. A tiny vibrating nanorod could do the same job in less space, using much less power.

It's really the prospect of nanorobots that gives the public nightmares. In December 1959, at the California Institute of Technology, the physicist Richard Feynman gave a prophetic lecture on manipulating and controlling things on a small scale, "There's Plenty of Room at the

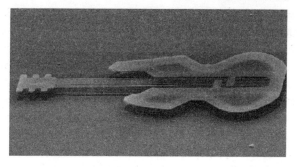

The Fender Stratocaster nanoguitar

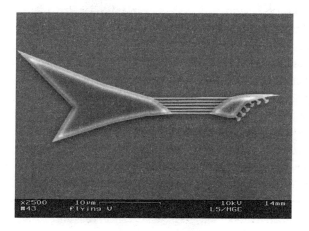

The 'playable' Gibson Flying V nanoguitar

Bottom." He envisioned nanomachines capable of building atomically precise products. At the time, scientists had succeeded in constructing electric motors the size of a small fingernail and had created a device capable of writing the Lord's Prayer on the head of a pin. Yet Feynman dismissed such advances as "the most primitive, halting step" on the road to miniaturization, foreseeing a day when scientists would be able to write the entire *Encyclopaedia Britannica* on the head of a pin. All the scientists had to do was reduce the size of the text 25,000 times. The necessary instruments and manufacturing techniques didn't exist when Feynman gave his lecture, but he was confident that they would not be long in coming.

As evidence for the possibility, Feynman pointed to biology, which is teeming with examples of "writing" information on a very small scale. Atoms are the building blocks of nature. The human body is composed of billions of living cells, which store all the information needed to coordinate all the functions of a complex organism. Cells are nature's nanomachines, active entities capable of manufacturing various substances, such as enzymes and hormones, and of storing information. For Feynman, this was proof of the possibility of building small but movable machines, computing at the quantum scale, arranging individual atoms, and perhaps eventually even performing internal surgery with miniature movable devices.

In *Prey*, Crichton's fictional Xymos Corporation is pushing its new medical imaging technology, which involves millions of miniature cameras the size of red blood cells being injected directly into the bloodstream with a syringe. The cameras then organize themselves into a networked "swarm." The idea is that a single camera the size of a molecule wouldn't be able to collect enough light to register a useful image. But millions of cameras linked together and operating simultaneously could produce a very clear composite image, much as the human eye creates an image from its rods and cone cells.

The Xymos system is fictional, but scientists do envision a day when patients may be able to drink a fluid containing millions of nanobots programmed to attack and reconstruct cancer cells and viruses to make them harmless. In fact, a research collaboration between Rice University and a company called Nanospectra has found that nanoparticles covered in gold will pool together in cancerous tumors when they are injected directly into the bloodstream. The nanoshells are then hit with an infrared (IR) laser beam. Normally IR light passes harmlessly through living tissue, but the gold-covered nanoparticles actually heat up sufficiently to burn away the tumors. Scientists hope to one day use nanobots to perform delicate operations, without leaving the scars resulting from conventional surgery. And while a swarm of nanobots isn't likely to be able to take on the appearance of an actual human being, as happens in *Prey*, it's conceivable that nanobots could be programmed to rearrange a person's atoms for cosmetic surgery—altering ears, nose, or eye color, for example.

Even more unsettling to the public mind is "molecular manufacturing." Early in the twentieth century, Henry Ford revolutionized manufacturing when he built the first automotive assembly line along the Rouge River in Michigan. At the time, bigger meant better. Today, the trend is toward making everything smaller, from laptops to cell phones to Apple's hugely popular iPod, a 6-ounce device capable of storing 10,000 songs. Nanotechnologists dream of a day when consumers will be able to order custom-designed goods, which will be pieced together, atom by atom, into a pattern that produces the desired structure, whether it be a baseball, an iPod, or an SUV. Scientists at IBM demonstrated the ability to manipulate atoms in 1990, when they were able to "grab" single xenon atoms and arrange them to spell out "IBM"—the world's smallest corporate logo.

The next step is to build nanoscale machines called "assemblers" capable of manipulating atoms and molecules into useful patterns. In 1959, Feynman suggested scaling down macroscale machines; but today, scientists believe that the most efficient route is to let the machines build themselves, a technique known as self-assembly. A single assembler working alone would take thousands of years to build a recognizable product one atom at a time, so we would need trillions of them, all working together, to match current manufacturing rates. The easiest way to build trillions of assemblers quickly is to make them self-replicating.

It's these types of tiny self-replicating nanoassemblers that wreak havoc in *Prey*. A mother who works at Xymos inadvertently, exposes her infant daughter, Amanda, to "gamma assemblers": tiny nanobots smaller than dust mites designed to cut carbon substrates to make finished molecules. Faced with a more pliable substance like skin, the gamma assemblers merely pinch it, causing a mysterious rash that is really bruising. Amanda is covered by a dusty coat of biting particles. Fortunately, she is instantly cured when an MRI machine's magnetic field yanks the nanobots away from her, although uniform bruising persists for a few days all over her body.

The public, alas, seems to struggle with separating fact from fiction. The Nobel laureate Richard Smalley, a chemistry professor at Rice University who helped pioneer nanotechnology, was deeply disturbed when he lectured to a group of 700 middle and high school students in Texas in 2003. The students were asked to write essays about nanotechnology, and Smalley read the 30 best submissions. He said, "Nearly half assumed that self-replicating nanobots were possible, and most were deeply worried about what would happen in the future if these nanobots spread around the world." In his view, Crichton, Drexler, and other writers are guilty of fear-mongering. He doesn't think nanoassemblers are physically possible, an opinion that led to a tense exchange with Drexler in *Chemical and Engineering News.*

Most nanoscientists agree that even if such nanobots are possible, practical production is decades away. "The idea that they're going to take over biological systems is absolutely silly," Jim Thomas, a fellow of the Royal Society University, told *Time Europe* in 2004. Drexler himself has backpedaled from his original alarm of 1986, claiming that he had raised the possibility as a worst-case scenario because he was worried that excitement over nanotechnology would overshadow the inherent risks. He still maintains that the "gray goo" scenario is "well within the realm of physical law," but he has admitted that he had "underestimated the popularity of depictions of swarms of tiny nanobugs in science fiction and popular culture."

Scientists may be tempted to snicker at Prince Charles's comments, but he is hardly alone in his concern about potentially harmful consequences from nanotechnology. In 2003, the Action Group on Erosion, Technology, and Concentration (ETC) in the United Kingdom called for an outright ban on all nanotechnology research until protocols for the safe handling of nanoparticles could be developed. In July 2004, an independent panel of British researchers recommended that cosmetics containing nanoparticles be banned from U.K. markets until they are proved safe for use on the skin.

Now it's the researchers' turn to be alarmed, with good reason. Smalley, like many others, is concerned that unfounded panic could slow or halt vital research in nanotechnology. Great Britain could be leading the charge. Some scientists argue that stopping research would itself be unethical, depriving society of the huge potential benefits offered by nanotechnology. Others are concerned that a backlash against nanotechnology would dismantle the potential trillion-dollar industry in its infancy—even if no environmental or health threats ever materialize. For Vicki Colvin, Smalley's colleague at Rice University, that prospect "is all too real, and just as frightening as anything a sci-fi author could imagine."

31

Contemplating Chaos

Circa January 1961
Lorenz and the butterfly effect

Nature's great book is written in mathematical symbols.
—Galileo Galilei

In Michael Crichton's best-selling novel (and the subsequent film) *Jurassic Park*, Ian Malcolm, a maverick mathematician—cursed with "a deplorable excess of personality"—draws on chaos theory to explain why an envisioned theme park featuring cloned dinosaurs is doomed to tragic failure, despite all efforts to keep the system perfectly controlled and predictable. Asked if perhaps he should reserve judgment until he's actually seen the island, Malcolm responds, "The details don't matter. Theory tells me that the island will quickly proceed to behave in unpredictable fashion."

To the average person, the concept of chaos brings to mind images of complete randomness. Yet to scientists, it denotes systems that are so sensitive to initial conditions that their output appears random, obscuring their underlying internal rules of order: the stock market, rioting crowds, brain waves during an epileptic seizure, or the weather. In a chaotic system, tiny effects are amplified through repetition until the system goes critical—an effect memorably illustrated by the carnage inflicted by Crichton's rampaging dinosaurs. This seemingly paradoxical concept was born when a mathematician turned meteorologist named Edward Lorenz made a serendipitous discovery that spawned the modern field of chaos theory, and changed forever the way we look at nonlinear systems like the weather.

Even as a boy in West Hartford, Connecticut, Lorenz was fascinated by the weather. After graduating from college, he planned to go into mathematics, but World War II intervened. He ended up serving as a weather forecaster in the Army Air Corps and enjoyed it so much that when the war ended, he decided to stick with meteorology.

Lorenz was particularly intrigued by weather prediction, which in those days was still largely intuitive guesswork. With the advent of computers, he saw an opportunity to combine mathematics and meteorology. He set out to construct a mathematical model of the weather using a set of differential equations representing changes in temperature, pressure, wind velocity, and the like. By late 1960, Lorenz had managed to create a skeleton of a weather system. He kept a continuous simulation running on his computer, which would produce a day's worth of virtual weather every minute. The system was quite successful at producing data that resembled naturally occurring weather patterns—nothing ever happened the same way twice, but there was clearly an underlying order.

Any hopes Lorenz had harbored about being able to make long-range weather forecasts—a major focus of his research—were dashed on a wintry day in 1961. He decided to take a closer look at one particular sequence in his simulation, and rather than start all over from the beginning, he began at a middle point, typing in the parameters of the earlier run straight from a computer printout. Then it was time for a much-deserved coffee break. But in his absence, something strange occurred. He returned to his office and found that instead of duplicating the earlier run exactly, the virtual weather diverged from the previous pattern so rapidly that within just a few virtual "months," all resemblance between the two had disappeared.

At first Lorenz assumed that a vacuum tube had gone bad in his computer, which was extremely slow and crude by today's standards. Instead the problem lay in the numbers he had entered to set the initial conditions. Six decimal places were stored in the computer's memory, but only three appeared on the printout. Lorenz had entered the shorter numbers. It shouldn't have made a difference in the outcome, since the discrepancy was only 1 part in 1000.

There was little reason to think otherwise. Scientists are often taught that small initial perturbations lead to small changes in behavior in any given physical system, and even today, temperature is not routinely measured within 1 part in 1000. Lorenz's computer program was designed so that the "weather" would unfold exactly the same way each time from any given starting point. A slightly different starting point would cause the weather to unfold in a slightly different way. Lorenz figured that a small numerical variation was similar to a small puff of wind, unlikely to have significant impact on important, large-scale features of the weather.

And yet that tiny discrepancy had changed everything. Today, this phenomenon is known as sensitive dependence on initial conditions. Lorenz called his discovery the "butterfly effect": the nonlinear equations that govern the weather have such incredible sensitivity to initial conditions that a butterfly flapping its wings in Brazil could theoretically trigger a tornado in Texas. And, as in Crichton's fictional theme park, Lorenz sadly concluded that it was impossible to achieve accurate, long-term weather forecasts, since at any time a slight variation in weather conditions could drastically change any forecast. Time has proved him right. Since that momentous discovery, computer modeling has advanced dramatically, transforming weather prediction from a largely intuitive art into a finely honed science. Yet forecasters can still only speculate about conditions beyond a few days at most.

This is also true of the TV reality series, *Survivor*, which might be viewed as a chaotic system of human relationships—if only to justify the viewer's own addiction. Its host Jeff Probst once remarked that even after several seasons, no clear pattern for a successful game-winning strategy had emerged. This is partly because human beings are innately unpredictable; alliances are formed and broken over the course of the game, and players can inexplicably change their minds at the last minute. Nor are the organizers above throwing the occasional curveball to keep the players on their toes. In short, the course of the game can turn on a dime—part of the show's lasting appeal. Mix all those variables together over 39 days, and correctly predicting the winner at the outset of each game is well-nigh impossible.

In the paranormal thriller *The Butterfly Effect* (2004), the beleaguered hero has the ability to turn back time and alter history. He discovers that changing just one thing—saving his girlfriend's life—changes everything, and not for the better. It's a theme that resounds throughout modern pop culture. In the holiday favorite *It's a Wonderful Life* (1946), George Bailey finds that the simple life of an outwardly unremarkable man in a small town can profoundly affect the lives of his friends and neighbors. And the plot of one of the most popular episodes of *Buffy the Vampire Slayer* centers on an ill-fated wish by the spoiled high school princess Cordelia that Buffy had never come to Sunnydale. The wish brings about a nightmarish alternative reality in which vampires rule the town, humans are chained and caged, and a powerful vampire known as the Master conceives of a diabolical plan for a "factory" to efficiently mass-produce fresh human blood. All of these are examples of the "butterfly effect" in action.

Of course, scientists had long observed such behavior—random fluctuations arising from what should have been a completely deterministic set of equations—but had discarded them as simply errors in calculation. Lorenz was the first to recognize this erratic behavior as something other than error; what he saw was a hidden underlying order, born out of randomness.

The mathematical offspring of chaos theory is fractal geometry. Like chaotic systems, fractals may appear haphazard at first glance, yet each one is composed of a single geometric pattern repeated thousands of times at different magnifications, like Russian dolls nested within one another. A fractal pattern is what is left behind by chaotic activity, rather like fossilized dinosaur tracks that enable us to chart their movement. If a hurricane is a chaotic system, then the wreckage strewn in its path is its fractal pattern.

Some fractal patterns exist only in mathematical theory, but others provide useful models for the irregular yet patterned shapes found in nature—the branchings of rivers and trees, for instance. Mathematicians tend to rank fractal dimensions on a series of scales between 0 and 3. One-dimensional fractals (such as a segmented line) typically rank between 0.1 and 0.9; two-dimensional fractals (such as a shadow cast by a cloud) between 1.1 and 1.9; and three-dimensional fractals (such as a mountain) between 2.1 and 2.9. Most natural objects, when analyzed in two dimensions, rank between 1.2 and 1.6. Actually, the specific dimensional numbers are less important than the general idea: a flat, grassy plain or coastline is less "busy"—and hence has a lower fractal dimension—than, say, a thick copse of trees.

Not content with being confined solely to the worlds of math and science, chaos and fractal geometry have made their way into art as well. Richard Voss, a researcher at Florida Atlantic University's Center for Complex Systems, has analyzed ancient Chinese landscape paintings from various historical periods at the Metropolitan Museum of Art in New York. He determined the dimension of brushstrokes in each painting and found that those considered superior by art historians (the earlier landscapes) had fractal dimensions comparable to those of typical coastlines: 1.25 and 1.33.

Perhaps the best example of fractal patterns in art is the work of the splatter master Jackson Pollock (aka "Jack the Dripper"). The physicist Richard Taylor was on sabbatical in England several years ago, pursuing a master's degree in art history, when he realized that the same analysis used in his scientific research could be applied to Pollock's work. Once back in his laboratory, he took high-resolution photographs of 20 canvases by Pollock dating from 1943 to 1952, and scanned them into a computer. Then he divided the images into an electronic mesh of small boxes. Finally, he used the computer to assess and compare nearly 5 million drip patterns at different locations and magnifications in each painting.

The result: very clear fractal patterns were detected in all the paintings. Nor was this merely a coincidence. The fractal dimensions of the earlier paintings correspond closely to those found in nature; a painting of 1948 (rather unimaginatively called *Number 14*) for instance, has a fractal dimension of 1.45, similar to that of many coastlines. The later the paintings, the richer and more complex the patterns, and the higher the fractal dimension. *Blue Poles*, one of Pollock's last drip paintings—now valued at more than $30 million—had the highest fractal dimension of any that Taylor analyzed: 1.72. Taylor believes that Pollock was deliberately testing the limits of what the human eye would find aesthetically pleasing.

In literature, scholars have analyzed fractal repetition in the poetry of Wallace Stevens and Gerard Manley Hopkins, as well as the Bible and *Moby Dick*. The impact in music is even greater. Voss pioneered the concept of fractal repetition in music as a postdoc at IBM under Benoit

Mandelbrot, who is considered by many to be the "father of fractals." The term "fractal music" refers specifically to music composed wholly or in part using the same types of feedback processes used to create fractal images, although certain classical composers like Debussy and Mozart are known to have used mathematically based composition techniques.

Phil Thompson, a computer programmer based in London, caused a sensation in 1997 when BBC Radio 4 broadcast a piece of his computer-generated fractal music and received a hugely positive response from listeners. (The piece—"A Season in Hell"—is based on the work of the French poet Arthur Rimbaud, who believed that art, music, science, and religion would one day be united.) Thompson designed a computer program for generating fractal music, called Gingerbread, and went on to release a much acclaimed CD, *Organised Chaos*, inspiring (for better or worse) a proliferation of composers of computer-generated fractal music. Programs like Gingerbread make musical composition accessible even to those who, like Thompson, don't happen to play an instrument. Instead, users simply click on a particular aspect of a fractal image, and the program translates the underlying mathematical formula into notes on a scale.

While this might sound suspiciously like music on demand accord-ing to a rigid mathematical structure, Thompson has staunchly defended his creation, insisting that the very nature of chaos dynam-ics ensures that the process is just as flexible and fluid as human creativity. Since there are numerous, perhaps infinite, ways of map-ping the numerical outputs to musical parameters, the choice of a spe-cific mapping has a significant impact on the resulting composition. In fact, it's impossible to determine beforehand exactly what the piece will turn out to be.

Why are fractal patterns so pervasive? Human beings might just be wired that way. Studies in perceptual psychology have found that people clearly prefer fractal dimensions similar to those found in nature; subjects in a study published in *Nature* in March 2000 preferred fractal dimensions between 1.3 and 1.5 roughly 80 per cent of the time. That same predisposition seems to be at work in other mediums as well. Other studies have found that people prefer patterns that are neither too regular, like the test bars on a television channel, nor too random, like a snowy screen. They prefer the subtle variations on a recurring theme in, say, a Beethoven concerto, to the monotony of repeated scales or the cacophony of someone pounding on a keyboard. In fact, artists, architects, writers, and musicians may instinctively appeal to their audiences by mimicking the fractal patterns found in nature.

According to James Wise, an adjunct professor of environmental sciences at Washington State University who has collaborated on some of these studies, those preferences may date back to our earliest ancestors. On the African savanna, for instance, they could probably tell whether the grass was ruffled by the wind or by a stalking lion by tuning into variations in fractal dimensions in their environment. But in settings with high fractal dimensions—such as a densely branching rain forest—early humans would have been more vulnerable, and hence more uneasy. Wise theorizes that perhaps our appreciation of lower-dimension fractal patterns isn't so much about beauty as it is a latent survival instinct.

Numerous artists have—either consciously or unconsciously—sought to capture the fractal characteristics of nature in their work throughout human history, but Taylor maintains that Pollock is still unique. Not only are his drip paintings fractal in dimension, but an analysis of film shot in 1950 by Hans Namuth depicting the artist working indicates that Pollock moved around his canvases in chaotic motions to produce the fractal patterns. He circled the canvas, dripping and flinging paint in motions that seem both haphazard and perfectly controlled. Thus he didn't merely imitate what he saw in nature; he adopted its very mechanism: chaos dynamics.

All this suggests that Pollock's paintings might be worth their multimillion-dollar prices, Taylor says. "If someone asked, 'Can I have nature put onto a piece of canvas?' the best example there has ever been of that is 1948's *Number 14.*"

32

Kamikaze Cosmos

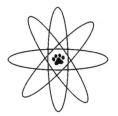

July 1963
Discovery of cosmic microwave background
radiation

A giant killer amoeba from outer space isn't something the average man on the street is likely to encounter, but for Captain James Kirk in *Star Trek*, it's just another day on the job. In a classic episode from the popular TV series, the *Enterprise* is sucked up by just such a creature, which feeds off the debris and other objects it encounters as it oozes randomly through space. Kirk and Scotty devise a clever plan to kill the creature with a magnetic bottle filled with antimatter siphoned from the starship's engine. They fire their makeshift bomb into the amoeba's nucleus, destroying the creature and freeing the *Enterprise.*

Diehard fans of the series no doubt have more than a passing familiarity with the *Enterprise's* unique fuel source, which supplies sufficient propulsion to boost the starship into its famous "warp drive." But others might not be aware that the concept of matter-antimatter propulsion is not just the stuff of science fiction. As with many technical aspects of the series, for the *Enterprise* propulsion system the creator of the series, Gene Roddenberry, drew on established scientific fact.

What is this mysterious stuff called antimatter? As its name implies, antimatter is the exact opposite of ordinary matter, made up of antiparticles instead of ordinary particles. Think of antiparticles as mirror images of regular particles. They are identical in mass to their regular counterparts. But just as looking in a mirror reverses left and right, the electrical charges of antiparticles are reversed. So an anti-electron would have a positive instead of a negative charge, and an antiproton would have a negative instead of a positive charge. When antimatter meets matter, the result is an explosion. Both particles are annihilated in the process, and their combined masses are converted into pure energy—electromagnetic radiation that spreads outward at the speed of light. The effect is ably illustrated in the feature film *Star Trek III: The Search for Spock*. After surrendering his ship to the Klingons, Kirk himself sabotages it, programming the computer to mix matter and antimatter indiscriminately, causing a huge explosion and destroying the *Enterprise*.

As far as we know, antimatter doesn't exist naturally in the universe. But scientists believe that 10 billionths of a second after the big bang, when the observable universe was about the size of the average living room, there was an abundance of antimatter. The nascent universe was incredibly hot—tens of billions of degrees—and infinitely dense, so that energy and mass were virtually interchangeable. It was an extremely violent environment—a Thunderdome for subatomic particles. New particles and antiparticles were continually being created and hurling themselves, kamikaze-like, at their nearest polar opposites, thereby annihilating both matter and antimatter back into energy, in a great cosmic war of attrition. Matter won.

In such a war, if both sides always have equal numbers of soldiers, neither can ever claim victory. But for reasons that continue to elude scientists, at some point in those first few fractions of a second, a small surplus of matter appeared. Even that tiny imbalance was sufficient to wipe out all the antimatter in the universe in about 1 second. As the universe expanded, the temperature began dropping rapidly, until it was too low to create new pairs to replenish the "armies." Only a smattering of "leftover" particles of matter survived the process—everything else had been annihilated and its mass emitted as radiation. Those bits and pieces make up the stars, planets, asteroids, and just about every other observable object in the universe.

Who dreamed up this crazy notion? Once again, we have Albert Einstein to thank. Remember that according to $E = mc^2$ if a body emits a certain amount of energy, then the mass of that body must decrease by the same amount. When Einstein first calculated this formula, scientists assumed that a particle's energy must always be a positive number. That idea changed in 1928, when the British physicist Paul Dirac speculated that either particles or antiparticles could exist, each particle with the same mass as its counterpart, but with the electrical charge reversed. In fact, he proposed the existence of a completely new universe made entirely out of antimatter. Soon a race was on to find some experimental means of proving the existence of antiparticles. It would be won by a young postdoc at CalTech named Carl Anderson, who was studying the nature of cosmic rays.

Cosmic rays are highly energetic subatomic particles traveling through space at nearly the speed of light. Many originate in the sun, but others come from objects outside our solar system. When these rays strike the Earth's atmosphere, they break apart into showers of other particles. Anderson used a cloud chamber to detect these secondary particle showers in hopes of determining what they were. A cloud chamber is a short cylinder with glass plates on either end. It is filled with gas saturated with water vapor, much like what happens when clouds become saturated with moisture that evaporates from the Earth's surface. Whenever a charged particle from cosmic rays passes through the chamber, it leaves a trail of water droplets that can be captured on film. The density of the droplets tells scientists how much charge is produced, and whether that charge is positive or negative. This in turn enables them to decide what kind of particle passed through the chamber.

Anderson's photographs surprised everyone: cosmic rays clearly produced showers of both positively and negatively charged particles. At first Anderson thought they were electrons and protons, but they all had the same droplet density—whereas a proton's would have been heavier than an electron's. Then he surmised that the positively charged particles were electrons traveling in the opposite direction. So he placed a lead plate in the chamber.

Particles that passed through the plate would emerge from the other side at a lower energy than when they started, so Anderson could determine the direction of travel. In August 1932, Anderson recorded a historic photograph of a positively charged particle passing through the lead plate in the cloud chamber. It was neither a proton nor an electron traveling backward; it was something new. Despite initial skepticism from other scientists, his result was confirmed the following year. His discovery won Anderson a Nobel Prize in physics in 1936, at the age of 31—he was one of the youngest people to be so honored.

Anderson's antiparticle was named a positron. More than 20 years later, scientists at the University of California, Berkeley, produced the first antiprotons: whereas the positron is in essence an antielectron, with a positive charge instead of a negative charge, antiprotons have a negative charge instead of the usual positive one. In 1995, researchers at CERN used a machine called the low-energy antiproton ring (LEAR) to slow down rather than speed up antiprotons. By doing so, they were able to pair positrons and antiprotons to produce nine hydrogen antiatoms, although these lasted only 40 nanoseconds. Within three years they were able to produce as many as 2000 anti-hydrogen atoms per hour, but this was not nearly sufficient to propel a starship like the *Enterprise*, in case anyone was excitedly planning an interstellar vacation.

Anderson's photograph of a positron

Still, antimatter propulsion systems aren't entirely in the realm of science fiction. The "Star Wars" strategic defense initiative in the 1980s included several projects that proposed using antimatter as rocket fuel, and as recently as October 2000, scientists at NASA announced early designs for an antimatter engine for future missions to Mars. Antimatter is an ideal rocket fuel because all the mass in matter-antimatter collisions is converted into energy. Matter–antimatter reactions produce 10 million times the energy produced by conventional chemical reactions such as the hydrogen and oxygen combustion used to fuel the space shuttle. They are 1000 times more powerful than the nuclear fission produced at a nuclear power plant, or by the atomic bombs dropped on Hiroshima and Nagasaki. And they are 300 times more powerful than the energy released by nuclear fusion.

If it weren't so difficult and expensive to produce in bulk, antimatter would be the most efficient fuel known to man. But the only way to produce antimatter is in particle accelerators (popularly known as atom smashers): enormous circular tunnels lined with powerful magnets that propel atoms around and around until they approach the speed of light. Then they slam into a target, creating a shower of particles. Some of those are antiparticles, which are separated out by the magnetic field. Even the most powerful atom smashers produce only minute amounts of antiprotons each year—as little as 1 trillionth of a gram, which would light a 100-watt bulb for barely 3 seconds. It would take tons of antimatter to fuel a trip to even the nearest stars.

Should an ample supply of antimatter be found, a secure means of storage must then be devised. The antimatter must be kept separate from the matter until the spacecraft needs more power; otherwise, the two would annihilate each other uncontrollably, with no means of harnessing the energy—and the resulting explosion would most likely destroy the spacecraft. That was the gist of yet another episode of *Star Trek*, in which the *Enterprise's* magnetic control system is sabotaged. The problem is discovered mere moments before the matter and antimatter cause a massive blast of energy, capable of annihilating the starship along with the crew—as happened in the movie *Star Trek III*.

Astronomers can still hear the echoes of the great cosmic battle between matter and antimatter in the radiation produced by all those annihilations: the cosmic microwave background radiation, discovered by physicists quite by accident in 1960. At the time, there were two prevailing theories about the origin of the universe. The more popular was the steady-state theory, which held that the universe was unchanging and would remain so forever. The big bang theory was more controversial. It sought to incorporate the astronomer Edwin Hubble's discovery (made in 1929) that galaxies are moving away from one another. A few physicists argued that the separation between galaxies must have been smaller in the past. The universe had once been infinitely dense, and all matter had emerged in a cataclysmic explosion: hence the term "big bang." But there was a sticking point: if the theory was correct, there would have to be a cosmic afterglow—the remnant of all those kamikaze collisions between matter and antimatter. The collisions produced huge amounts of energy in the form of radiation, and when the universe cooled and the matter-antimatter annihilations were no longer taking place, the radiation remained. That background radiation cooled along with the expanding universe, until it reached a uniform temperature of about 2.73 kelvins (K).

In 1960, two scientists at Bell Laboratories, Arno Penzias and Robert Wilson, salvaged a 20-foot horn-shaped antenna in Holmdel, New Jersey, part of an obsolete satellite transmission system that Bell had developed years earlier. They wanted to use the antenna as a radio telescope to amplify and measure radio signals from the Milky Way and other galaxies. To do so, they had to eliminate all interference from their receiver: both radio and radar broadcasting, and noise from the receiver itself. But even after removing all known interference, Penzias and Wilson still found an annoying hissing noise in the background, like static. The noise was a uniform signal falling into the microwave frequency range, and, strangest of all, it appeared to be coming from all directions at once.

The two scientists checked every possible source of the excess radiation. The signal wasn't from urban interference, nor was it radiation from the Milky Way galaxy or other celestial radio sources. Then they decided that the problem might be due to the droppings of pigeons roosting in the antenna. So they contrived a pigeon trap to oust the birds, and spent hours removing pigeon dung from the contraption. But the strange hissing persisted. In desperation, they consulted with a colleague at nearby Princeton University, Robert Dicke, who had first predicted low-level background radiation if the big bang theory was correct.

Dicke realized the potential significance of the signal immediately, and rushed out to Bell Labs to investigate. He confirmed that the mysterious signal was indeed the cosmic microwave background radiation, and resignedly told his colleagues at Princeton, "We've been scooped." In 1978 Penzias and Wilson received the Nobel Prize in physics for their accidental discovery, prompting one colleague to joke that the pair had "looked for dung but found gold, which is just opposite of the experience of most of us." The giant radio antenna is now a national historic landmark. Even the lowly pigeon trap has found its way into posterity: it's part of a permanent exhibition at the Smithsonian Institute's National Air and Space Museum in Washington, D.C.

Thus far all the evidence indicates that the universe did indeed start with a bang, and that kamikaze collisions between matter and antimatter most likely gave rise to the great cosmic structures in the universe today—as well as the background hiss of microwave radiation. Who knows? Perhaps one day scientists will achieve a practical means of matter-antimatter propulsion to enable a future generation of astronauts to boldly go where only the fictional *Enterprise* has dared to venture before.

33

When Particles Collide

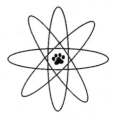

April 1994
Discovery of the top quark

Many a bride has quaked at the prospect of introducing her future in-laws to her parents. This is especially daunting for Toula Portokalos, the curvacious heroine of the sleeper hit film *My Big Fat Greek Wedding* (2001). Toula has always cringed inwardly at her Greek family's pronounced eccentricities. She had to go to Greek school instead of Girl Scouts, and her school lunches consisted of moussaka (called "moose ca-ca" by her jeering classmates) rather than peanut butter and jelly sandwiches. She grew up in the only house on the block with white pillars reminiscent of the Parthenon, and a Greek flag painted on the garage door. And her proudly patriotic father constantly regales friends and neighbors with examples of the superiority of Greek culture. Engaged against her father's wishes to Ian Miller, a preppie high school teacher, Toula hopes for a quiet, low-key dinner to introduce the two sets of parents. Instead, the reticent, WASP-ish Millers arrive—traditional Bundt cake in hand—to find the entire extended Portokalos clan gathered on the front lawn, roasting a pig on a spit. The Millers are understandably dazed at encountering such a big, boisterous bunch of relatives with strange-sounding names and even stranger customs.

The Millers' confusion is similar to the befuddlement most non-physicists feel when first encountering the large, unruly "family" of elementary particles. Subatomic particles arise from their own culture clash: the collision of atoms inside huge machines known as particle accelerators, or atom smashers. Atom smashers are long underground tunnels that speed up beams of particles to very near the speed of light and then smash them into target atoms. The cathode-ray tube inside a television is a miniature atom smasher. The cathode emits electrons into the vacuum tube, where they speed up and smash into phosphor molecules coating the TV screen, causing them to glow. The result is a lighted spot on the screen. Particle accelerators work in much the same way. Most of the particles only glance off each other, but every now and then, two particles collide head-on, producing a shower of smaller subatomic particles.

The notion of smashing atoms calls to mind the explosive power of an atomic bomb. In reality, when particles collide, they release only a small amount of energy—barely enough to propel a mosquito. But inside the confines of a particle accelerator, these collisions are strong enough to crack pieces of atomic nuclei into their most fundamental parts. We saw in Chapter 32 how collisions between matter and antimatter just after the big bang gave rise to all the different types of particles in the universe. Atom smashers re-create that volatile environment on a much smaller, controlled scale. In particle accelerators, physicists can create tiny fireballs to simulate the conditions that existed mere fractions of a second after the big bang.

Smashing atoms is the best method scientists have found to test the theoretical predictions of the "standard model," which describes the basic building blocks of the universe and how matter evolved. Those blocks can be divided into two basic clans: fermions and bosons. Fermions make up all the matter in the universe, and include leptons and quarks. Leptons are particles, such as electrons and neutrinos, that are not involved with holding the atomic nucleus together. Their job is to help matter change through nuclear decay into other particles and chemical elements, using the weak nuclear force. Quarks help hold the nucleus together through the strong nuclear force. The physicist Murray Gell-Mann is credited with adopting the term "quark." It comes from a line in James Joyce's *Finnegans Wake:* "Three quarks for Muster Mark."

Bosons are the ties that bind the other particles together, in much the same way that love, loyalty, and generous shots of ouzo bind

Toula's family into a tight social unit. Bosons pass from one particle to another, giving rise to forces. A force is basically a "push" or a "pull." Let's say that Ian and Toula are playing catch on roller skates. Every action has an equal and opposite reaction. So if Toula throws the ball and Ian catches it, they will be pushed in opposite directions by the force that results from the exchange. That's how bosons work. The ball is the boson—the carrier of the force between Ian and Toula, who represent two different particles—and the repulsion is the force itself. We can see the effects of the resulting force, but not the exchange that gives rise to it. There are four force-related "gauge bosons." The gluon is associated with the strong nuclear force: it "glues" an atom's nucleus together. The photon carries the electromagnetic force, which gives rise to light. The W and Z bosons carry the weak nuclear force and give rise to different types of nuclear decay. Scientists have yet to incorporate gravity into the standard model, so there is an unsightly gaping hole in an otherwise elegantly symmetrical system.

Those are just the main players, the most fundamental bits in the subatomic zoo. There's also a host of supporting characters within those two basic clans: metaphorical aunts, uncles, nieces, nephews, cousins, and even second cousins. For example, when quarks combine, they form hadrons—protons, neutrons, and pions—and baryons. Add in mesons (quark-antiquark pairs) and all the different types of antimatter, and there are 57 different subatomic particles that physicists have confirmed so far. Only the Portokalos family wouldn't find that a bit excessive.

Things were not always so complicated. "Atom" means "indivisible," and in the nineteenth century, the atom was believed to be the smallest unit of matter. It was something of a surprise when physicists realized in the early twentieth century that the atom was made of even smaller particles: the proton, neutron, and electron. In the 1930s, they found that cosmic rays—fast-moving protons from outer space—would produce sprays of smaller particles if they collided with atoms of lead. Clearly the atomic nucleus was made of much smaller pieces than protons and neutrons. Quarks are now believed to be the smallest of all.

Initially the only way to get fast-moving particles to collide was to conduct experiments on mountaintops, where cosmic rays were more abundant. First invented in the 1920s, atom smashers finally became powerful enough in the 1950s to enable physicists to bring their experiments down to earth. After that, discoveries of new particles came fast and furiously. By 1985, all the subatomic particles predicted by the standard model had been found, except one: the top quark.

Because it's so much heavier than the other quarks (as heavy as a gold atom), the top was particularly difficult to find. It required colliding particles at much higher speeds in order to achieve the necessary high energies. It's also the most unstable, with a lifetime of mere billionths of a second. Particle physicists pinned their hopes for detecting the top quark on an atom smasher at Fermi National Laboratory (Fermilab) in Illinois, called the Tevatron. It's a huge facility: the circular track of the Tevatron is 4 miles long. And it's very much an international affair. More than 2000 physicists from all over the world are running experiments there at any given time. Roughly 900 of them were part of the two primary collaborations devoted to finding the top quark.

The ultimate goal of all accelerator experiments is to learn what kinds of particles make up the atom, and what forces hold it together. Suppose that Toula's father dropped the family television off the roof, smashing it to bits, and then tried to figure out how it worked by examining all the pieces. That's what particle physicists do. Fortunately, the particles themselves can help us to "see" what's inside an atom. For example, much as the reserved Miller family serves as a foil to the boisterous Portokalos clan, leptons can be used to probe quarks because they don't interact directly with the strong nuclear force. When they collide head-on with an atomic nucleus, they bounce off and scatter. Physicists can tell what's in the nucleus by studying the patterns of the scattering.

Physicists are able to recognize the particles produced in these collisions by the electronic signatures they leave behind. These "signatures" are nuclear decay patterns. Quarks exist for only fractions of a second before they decay into other secondary particles. Since each quark has many different ways of decaying, there are several possible signatures, and each must be carefully examined to determine which particles were present at the time of the collision. Decay patterns are like branching generations in the family tree, and every bit as complicated. Keeping all the different patterns straight is as difficult as keeping track of the jumble of Portokalos relatives.

In fact, to decipher all the signatures, physicists use a handy "cheat sheet" compiled by some physicists at Berkeley called the Particle Data Group. They bring together everything that's currently known about the existence and properties of the particles. It's somewhat akin to trying to make sense of the hastily scrawled signatures of the wedding guests in Toula's guest book. There are the usual chatty aunts, laconic uncles, and bickering cousins that typically show up for big family events. But every now and then, a more rare and interesting signature turns up: a long-lost brother, perhaps, or Mr. Portokalos's senile eccentric mother, who wanders the neighborhood in her widow's weeds, cursing in her native Greek, convinced that the Turks are trying to murder her family. The top quark is the rarest of all, equivalent to having a recluse like J. D. Salinger or Greta Garbo appear and sign the guest book. Moreover, a mischievous cousin might forge a signature, or bad handwriting could mask a guest's true identity.

Let's look at just one possible "signature event" for the top quark. First, a proton and an antiproton collide and produce a top quark and an antitop particle. These instantly decay into two weak-force (W) bosons and two bottom quarks. One of the "offspring" bosons turns into a muon and a neutrino (both members of the lepton family), while the other decays into up and down quarks. The two bottom quarks decay into two jets of particles, as do the up and down quarks. So the end result, or "signature," of the collision is a muon, a neutrino, and four jets. "Jets" appear because quarks can't exist in isolation; they must be bound inside hadrons. Whenever a quark is produced in a collision, it goes flying out of its host hadron. But before it can escape completely, it is surrounded by a spray of hadrons all traveling in more or less the same direction. By studying the jet spray, scientists can tell what kind of quark produced it.

Confusing, isn't it? Subatomic particles breed like rabbits. Figuring out which "parent" particles give rise to specific decay patterns can be as difficult as tracing your family genealogy back to Adam and Eve. And it all happens in fractions of a second. That's why detectors are needed to keep track of what's happening and make sense of all the data. Scientists use many different types of detectors in combination, each with a specific function. The detectors act as a filter, picking out just those signals that indicate a likely top event, out of the tens of thousands of signals created every millionth of a second inside the accelerator. This still amounts to almost 100 megabytes of data every second. The physicists working with the Tevatron were essentially looking for one particular subatomic needle in a field of haystacks. It was a long, laborious process of sifting through the tons of background noise and "false positives" that can mimic the top quark's signature.

They didn't find the top quark overnight. Fermilab waited an entire year to announce the discovery after pinpointing the first evidence in 1994. Ultimately, everything came down to statistics. The scientists at Fermilab expected to see roughly four top quark signature events after every run, but that was on average. Sometimes they saw two, sometimes six. So to make sure they were really seeing the top quark's signature, and not an imposter, they needed to find many more events than expected. Without a large enough "sample" of events, there was a strong probability of mistaking random background fluctuations for top signatures. "We discovered the top quark not in one lightning stroke, but over a long period of time, event by event," said Nick Hadley, one of the physicists involved in the project, when the discovery was finally announced. "No single piece of evidence, no matter how strong, was enough to let us claim a discovery. We couldn't be sure we had found the top quark until we had seen so many events with the right characteristics that there was almost no chance the statistics were fooling us into making a false claim." In the end, the probability that they had been fooled by background noise was less than 1 in 500,000.

Since then, the Tevatron has been upgraded and revamped to make it even more powerful. When the collaborators resumed collecting data in 1999, they found that the upgrades produced top quarks at 20 times the previous rate, and the detectors were much more efficient at identifying top quark signature events. A new atom smasher called the Large Hadron Collider (LHC), to begin operating in Switzerland in 2008, is expected to reach unprecedented high energies. Scientists hope that it will generate larger numbers of top quarks, and perhaps shed light on another as yet undiscovered particle: the "Higgs boson."

Physicists believe the Higgs boson carries the force that gives some of the subatomic particles their mass. It's a long shot—tantamount to a miracle in the minds of some physicists—but there may also be another boson called a "graviton" associated with the gravitational force. Finding it might fill the gaping hole in the standard model.

For all the Sturm und Drang that results from clashes of culture, or of subatomic particles, eventually the dust settles and things cool down to a comfortable equilibrium. Mr. Portokalos learns to accept the Millers, and they in turn embrace the Greek family's colorfully eccentric world, where a spritz of Windex can cure a host of ills, and spitting on the bride as she walks down the aisle is believed to bring good luck. The world of elementary particles is an equally strange and wonderful place. As different as the various particles may be, we need the quarks and leptons to make matter, and we need the bosons to forge the connections between them. As Mr. Portokalos so aptly puts it, there may be a difference between apples and oranges, but in the end, we are all fruit.

34

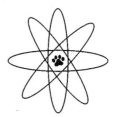

Checkmate

May 1997
Deep blue defeats Kasparov

An early scene in the sci-fi film *Blade Runner* (1982) features a psychological evaluation of a secretary in the sleek, expensive headquarters of Tyrel Corporation, a manufacturer of artificial life-forms. The examiner, Deckard, quizzes the woman for hours before arriving at a surprising conclusion: Rachel isn't human at all. She is a replicant: an android designed to mimic human beings in every way, except for emotional response.

In the director Ridley Scott's brooding vision of Los Angeles in 2019, replicants ("skin jobs" in street slang) are used as slave labor "off world," but their presence on Earth has been outlawed. Violators are "retired" by professional assassins called "blade runners"—Deckard's former line of work. He has been forced to return to active service to hunt down four rebel replicants. But Rachel is different: an advanced model fully equipped with manufactured memories, which make her sufficiently emotionally developed to fall in love with Deckard, and to be loved in return. Even she believes herself to be human, and she is devastated to discover otherwise.

In popular science fiction, robots and computer networks are always evolving intelligent consciousness, usually with disastrous consequences for the human race. In 1998 William Gibson wrote an episode for the *X-Files* in which a computer program becomes sentient and destroys its own creator, a theme also explored in the

Terminator films. Rebel humans in *The Matrix* are hunted down by "agents": sentient computer programs with ultrafast processing speeds that enable them to manipulate the virtual environment much more effectively than mere humans, whose brains are hardwired to believe that the laws of physics still apply. A modern-day Pygmalion would not create his perfect woman as a marble sculpture. He would construct her out of wires and circuit boards and program her to satisfy his every whim—a concept taken to sinister extremes in the 1970s in the horror film *The Stepford Wives* (inexplicably remade as a comedy in 2004).

In reality, however, artificial intelligence (AI) has proved to be more elusive. The concept dates back to Aristotle, who was the first to think about the rules governing human thought. In his treatise *Politics*, he justified slavery by arguing that the only way to dispense with the master–slave relationship would be if machines could work autonomously, using intelligent anticipation. More than 2000 years later, modern scientists have made great strides in building computers that can mimic strictly defined logical thought, but they still haven't cracked the code of human emotion and consciousness.

There are 100 billion nerve cells, called neurons, in the human brain, each of which can have up to 1000 connections. Neurons gather and transmit electrochemical signals, much like the logic gates and wires in a computer. But the similarity ends there. A computer's architecture is radically different from the neural structure of the brain. Most electronic systems are either analog (radios, TVs) or digital (microprocessors), but the brain is a combination of both. Take sensory perception as an example. If Deckard goes out hunting replicants in the rain—it is always raining in *Blade Runner*— he hears the raindrops, smells the wet pavement, and feels the rain on his skin. This is all generic analog input. His brain will digitally process this information and conclude that it is raining. So neurons make digital decisions, based on the collective analog data they receive as raw sensory input from a vast network of connections.

There are two prevailing schools of thought on AI. Proponents of strong AI believe that all human thought is essentially algorithmic and can be broken down into a set of mathematical operations. They believe that they will one day be able to replicate the human mind and create a genuinely self-conscious robot capable of both thinking and feeling—the stuff of classic science fiction. Proponents of weak AI believe that human thought and emotion can only be simulated by computers. A computer might seem intelligent, but having no sense of self or consciousness, it is not aware of what it is doing.

A professor of philosophy at the University of California, Berkeley, John Searle, illustrated this with his famous "Chinese room" parable. Place an English-speaking man inside an enclosed room with a rule-book that tells him how to respond to Chinese sentences that are pushed through a slot in the wall. To observers, he appears to be fluent in Chinese, because his responses are given so fluently. But the man is merely processing requests; he doesn't understand either the phrases or his responses. It's little more than pattern recognition. Similarly, Searle contends that a mere machine with a programmed set of responses cannot be considered truly intelligent—even if it proves to be particularly good at chess.

In 1770, the court of Empress Maria Theresa of Austria witnessed a seemingly miraculous phenomenon: a carved wooden human figure in elaborate Turkish dress that played tournament-level chess. The "Turk" defeated nearly all challengers. People marveled at this seemingly intelligent machine, but in reality, the Turk was an elaborate illusion created by the inventor Baron Wolfgang von Kempelen. Its base was a wooden cabinet containing an elaborate system of wheels and pulleys that concealed a human player, who moved the Turk's arms from inside the cabinet. Von Kempelen made a big show of opening doors and panels to prove that no one was concealed inside, and winding up the mechanism to distract audiences from the trick. Yet he never claimed that his invention was anything more than an illusion, although he guarded its secret vehemently. Some 15 different chess experts occupied the Turk during the 85 years it toured Europe and Russia, defeating such celebrated personages as Benjamin Franklin, Frederick the Great, Edgar Allan Poe, and Napoleon Bonaparte.

The Turk may have been the first (apparent) chess-playing automaton, but it wasn't the last: there were at least two others that toured fairgrounds into the early twentieth century. And because it is so soundly based on calculating logical progressions, chess proved to be an excellent fit for modern computers. The first chess computer

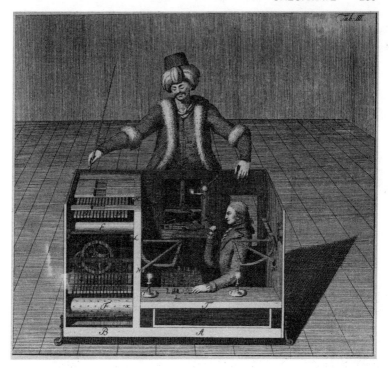

**Von Kempelen's
chess-playing Turk**

program was written in the late 1950s. Two decades later, the Machack IV computer became the first to play in a human chess tournament. By the time grand master Gary Kasparov first became world champion in 1985 at the age of 22, solid chess-playing machines were commonplace. Still, IBM's Deep Thought computer lost miserably to Kasparov in 1986.

Nearly 10 years later, a graduate student at Carnegie Mellon named Feng Hsiung Hsu developed a chess-playing computer called Chiptest. After earning his PhD, Hsu joined the research staff at IBM. He and his colleague Murray Campbell adapted his work on Chiptest as part of an effort to explore how to use parallel processing to solve complex computer problems. The Deep Blue project was born. Deep Blue was an impressive piece of machinery, capable of calculating 1 billion positions per second. Yet 97 per cent of the computer was constructed from components that could be purchased by the average consumer at Radio Shack. By February 1996, Hsu felt that his creation was ready to face the reigning world champion of chess, and challenged Kasparov to a match.

Kasparov was confident going into the match, but Deep Blue stunned the experts by winning the first game. It offered a pawn sacrifice early on to gain a lead in position: a common strategy among chess players, but a risky one, since the outcome is uncertain. The computer went on to recover the sacrificed pawn and win the game. Kasparov later told *Time* magazine that he was "stunned" by the computer's decision to sacrifice a pawn. "I had played a lot of computers, but had never experienced anything like this," he said. "I could feel a new kind of intelligence across the table."

A consummate competitor, Kasparov recovered his equanimity and ended up winning the match. He eventually defeated the computer by switching strategies in mid-game, since the computer did not so much think and strategize as react to its opponent's moves by analyzing all the possible moves and countermoves. "My overall thrust was to avoid giving the computer any concrete goal to calculate toward," Kasparov said. "So although I did see some signs of intelligence, it's a weird kind, an inefficient, inflexible kind that makes me think I have a few years left."

Kasparov had only about one year left. In May 1997, he faced off against the latest, improved iteration of Deep Blue in a rematch that made history. Early in game 6, Kasparov made a disastrous mistake, allowing Deep Blue to sacrifice a knight, obtain an overwhelming positional advantage, and go on to take Kasparov's queen in exchange for a rook and a bishop. Kasparov resigned the match after only 19 moves.

Credit for Deep Blue's success was given to the improvements Hsu had made after the first match. First and foremost was speed: faster processors gave the computer the ability to evaluate 200 million positions per second. Kasparov can examine only approximately three positions per second. And Deep Blue's knowledge of chess was significantly enhanced so that it could draw on vast resources of stored information, such as a database of opening games played by grand masters over the last 100 years. The extra computing power also allowed Deep Blue to adapt to new strategies as the game progressed—something it had been unable to do in the match won by Kasparov in 1996.

Deep Blue's triumph was heralded by some as a step forward in the march toward artificial intelligence, but others were more skeptical. "From a purely logical perspective, chess is remarkably easy," says Selmer Bringsjord, a professor of philosophy at Rensselaer Polytechnic Institute and a proponent of weak AI. "Invincible chess can theoretically be played by a mindless system, as long as it follows an algorithm that traces out the consequences of each possible move until either a mate or a draw position is found."

True AI should be able to learn from its mistakes and adapt to its environment. Bringsjord believes that it should also be creative—an idea based partly on his own experience writing a novel. "Processing speed is irrelevant to what's most impressive in human cognition," he says. "To sit down with a pen before a blank piece of paper and produce a play like Hamlet involves doing something that no computer, however fast, can pull off. Computers don't have, and therefore can't adopt, points of view. They don't have feelings; they have inner lives on a par with those of rocks. No amount of processing speed is ever going to surmount that obstacle."

With the scientist Dave Ferucci of IBM, Bringsjord created Brutus. 1, an artificial author programmed to create its own short fiction, generating stories of 500 words or less every few seconds, all based on the concept of betrayal—hence the computer's name. But Brutus. 1 is hardly a silicon Shakespeare, or even a silicon Danielle Steele. Bringsjord says that, far from demonstrating the creative powers of computers, his device exposes their limitations. In Ray Bradbury's short story "I Sing the Body Electric," three children who have lost their mother are raised by a sentient robot nanny. One of them asks whether she loves her charges. She responds that it depends on how he defines love. If love is defined as taking care of people and tending to their needs without complaint, then she must love them. She identifies love not by feeling, but by how it is defined through logic. That's how Bringsjord trained Brutus.1: he devised a formal mathematical definition of betrayal based on the plots and characters of famous authors. The machine was ignorant of other major literary themes, from unrequited love to revenge and jealousy, and it had no concept of character—or of any real emotion.

Bringsjord's latest project is designing synthetic characters, similar in concept to those used in the hugely popular SIMS computer games. One prototype is dubbed "E," since it will personify evil. To create it, Bringsjord is compiling a database of examples of evil characters gleaned from fiction, real life, and film, from Arnold Schwarzenegger's original Terminator to the repressive, controlling father who drives his son to suicide in *Dead Poets Society*. This will enable Bringsjord to create a clinical definition of evil that he can then translate into a mathematical algorithm. Eventually he plans to combine the Brutus project with the synthetic characters to create a computer than can spin stories containing both plot and characters. The end result may not be truly sentient AI, but like the chess-playing wooden Turk, it will be a very convincing illusion. "I want to see if I can engineer a system that will convince you that you're interacting with a human being, and not a synthetic character," he says. "We can't capture Shakespeare's imagination, but perhaps we can capture the structures that his imagination manipulated."

Creativity might be beyond a computer's reach, but Rodney Brooks, a scientist at MIT, believes that artificial emotion might one day be possible. He is the creator of Kismet, a robot that can respond to social cues; and in recent years he has turned his attention to building fleets, or "swarms," of robotic insects that work together to adapt to their environment. It's part of so-called "evolutionary robotics": creating machines that are digitally "bred" to evolve themselves. Swarm intelligence is the notion that complex behavior can arise when large numbers of individual agents each follow very simple rules. For example, ants follow the strongest pheromone trail left by other ants to find the most efficient route to a food source, through a process of trial and error. A chunk of the plot in Michael Crichton's novel *Prey* was inspired in part by an experiment in which a fleet of robotic predators were programmed to seek out "prey" to get their next energy boost. The mechanical "prey," in contrast, were programmed to "graze" on special light sources and to keep alert for potential predators. The respective robots evolved increasingly complex hunting and escape strategies as the swarms of robots accumulated more and more data (in the form of experience) on which to base their decisions.

Brooks believes that if a robot can adapt, it must be capable of evolving at least rudimentary emotions. That could have unforeseen consequences for Aristotle's vision of mechanical slave labor. In Karel Ĉapek's play *R.U.R.* (1920), robotic machine workers begin to chafe at servitude and overthrow their human creators when a scientist gives them emotions. The replicant slaves in *Blade Runner* are programmed to "die" after four years because they become much more difficult to control as they begin to develop emotions—and they, too, stage a bloody mutiny against their human "masters." We see emotional development firsthand in the chief rebel replicant, Roy. Designed to be a killing machine, he fulfills his lethal purpose for much of the film with ruthless efficiency. But he exhibits genuine grief at his comrades' deaths, and his final act is one of mercy: rescuing Deckard, his archenemy, who is dangling perilously from a rooftop.

Perhaps the true measure of human sentience is awareness of one's own mortality. Roy's primary motivation for returning to Earth is the most basic of human instincts: survival. He simply wants to live, only to discover that his "expiration date" is irreversible. Gazing at one last sunrise as life ebbs out of him, he mourns the fact that all the extraordinary things he has seen and done "will be lost in time, like tears in the rain." The game of chess at least offers the option of playing one's opponent to a draw. Roy ultimately understands—better than most humans—that life is a game we are all destined to lose, whether we are kings, paupers, chess grand masters, or killer androids from outer space. It is always check and mate. Should a real-life robot or computer ever have a similar insight, the quest for true AI will be over.

35

Much Ado About Nothing

December 1998
Discovery of the accelerating universe

A mysterious substance known only as "dust" pervades Philip Pullman's fantasy trilogy, *His Dark Materials*. In the first book, *The Golden Compass*, we learn that some years earlier, in an alternative universe, a Muscovite "experimental theologist" named Boris Rusakov discovered a new kind of elementary particle. Particles of this new type had no charge and seemed to cluster around human beings as if attracted to them. The existence of this particle didn't fit within the purview of the narrow ideology of the church, so naturally Rusakov was first suspected of demonic possession, then tortured, just in case he was lying. But ultimately the church was forced to accept that the mysterious particles did exist. And it concluded that "dust" was an emanation of the "dark principle" in the universe—the direct result of original sin.

A less fanciful interpretation is given in the second book, *The Subtle Knife*, which takes place in our "real" world. Mary Malone, a physicist at Oxford, is searching for something called dark matter. "We can see the stars and galaxies and the things that shine, but for it all to hang together and not fly apart, there needs to be a lot more of it—to make gravity work," she explains to Lyra, the plucky preadolescent heroine of Pullman's trilogy. Mary has discovered something she calls "shadow particles" that seem to fit the requirements for the dark matter. Lyra says that the "shadow particles" are the same thing as the dust particles found in her world.

"Dust" and "shadow particles" may both be figments of Pullman's prodigious imagination, but the existence of dark matter is a very real scientific conundrum. Compared with the universe in the first few seconds after the big bang, the universe we live in today is a vast, cold, and largely empty space, with a few clusters of galaxies scattered about, and nothing much in between. But to physicists, this great expanse of nothingness is filled with an invisible something. As Mary Malone aptly explains, the dark matter serves as a "gravitational glue" to keep galaxies from flying apart.

The problem is that no one can detect it. Astronomers rely on light to observe celestial bodies, but the dark matter neither emits nor reflects light. Its existence must be inferred. Physicists have no idea what the dark matter is: black holes, accumulations of hydrogen gas known as "brown dwarf" stars, or a mysterious type of shadow particle that passes right through normal matter and interacts with it only through gravity. And the mystery gets even weirder. Scientists believe that ordinary matter accounts for a mere 4 per cent of all matter in the universe, while dark matter accounts for 23 per cent. That leaves 73 per cent unaccounted for. So they have concluded that not only does nothing contain an invisible form of matter; it also has an equally strange form of invisible "dark energy." The exact nature of both is just as elusive as the nature of Pullman's dust.

It's the sort of baffling puzzle that scientists find endlessly fascinating, but why should the rest of us care about nothing? Determining the nature and source of all that dark stuff is directly related to how the universe will eventually end. It's the ultimate "big picture" question, even though we're talking about time scales of hundreds of billions of years. The cosmologist Michael Turner considers it "the most profound mystery in all of science."

Once upon a time, physicists believed that the cosmos was static and unchanging, a celestial clockwork mechanism that would run forever. When Albert Einstein was forming his theory of general relativity in 1917, his calculations indicated that the universe should be expanding. But this didn't fit with prevailing scientific opinion. So he introduced a mathematical "fudge factor" into his equations, known as the cosmological constant, or lambda. It implied the existence of a repulsive force pervading space that counteracts the gravitational attraction holding the galaxies together. This balanced out the "push" and "pull" so that the universe would indeed be static.

Perhaps Einstein should have trusted his instincts. Twelve years later, when the astronomer Edwin Hubble was studying distant galaxies, he noticed an intriguing effect in the light they emitted: it had a pronounced "shift" toward the red end of the electromagnetic spectrum. This is known as the "Doppler shift," after the Austrian physicist Christian Doppler, who in 1842 had noticed that sound will change its pitch with its speed relative to a listener. For instance, as a train passes, the pitch in the whistle changes from high to low. As the train approaches, the sound waves in the whistle are compressed, making the pitch higher than if the train were just sitting in the station. And as the train moves away, the sound waves are stretched, lowering the whistle's pitch. The faster the train is moving, the greater the change in pitch. The same thing happens with light. When a light source is moving towards an observer, the wavelength of its emitted light compresses and shifts to the blue end of the spectrum. When it is moving away from the observer, the wavelength stretches, and the light shifts to the red end of the spectrum.

Light from the most distant galaxies has traveled billions of years, giving astronomers a snapshot of the universe at a fraction of its present age. So when Hubble observed the red shift coming from distant galaxies, he reasoned that this could be happening only if the light was traveling across space that is expanding. Picture an image drawn on the surface of a balloon. As the balloon is inflated, its surface stretches, and the image stretches with it. That's what happens to light waves as they travel through expanding space: their wavelength stretches and shifts to the red end of the spectrum. The longer the light has been traveling, the more time there has been for space to expand, and hence the greater the red shift of the light's wavelength. Hubble's conclusion was inescapable. Einstein's original equations had been correct, and there was no need for a cosmological constant. The cosmos was indeed still expanding. Einstein denounced lambda as his "greatest blunder" when he heard of Hubble's discovery.

This changed the big picture of how the universe will end. If the universe was still expanding, scientists reasoned, eventually the attractive force of gravity would slow down the rate of expansion. They spent the next 70 years trying to measure that rate. If they knew how the rate of expansion was changing over time, they could deduce the shape of the universe. And its shape was believed to determine its fate. Matter curves space and time around it and gives rise to what we recognize as gravity. The more matter there is, the stronger the pull of gravity, and the more space will curve—making it more likely that the current expansion will halt and the universe will collapse back in on itself in an antithesis to the big bang called the "big crunch." If there's not enough matter, the pull of gravity will gradually weaken as galaxies and other celestial objects move farther apart, and the universe will expand forever with essentially no end. A flat universe, with just the right balance of matter, means that the expansion will slow down indefinitely, without a recollapse.

All the evidence to date has pointed to a flat universe. But just as physicists were getting comfortable with the fact that the universe would slow its expansion forever, the story took an unexpected turn. In 1998, two separate teams of physicists measured the change in the universe's expansion rate, using distant supernovas as mileposts. Because they are among the brightest objects in the universe, these exploding stars can help astronomers determine distances in space. By matching up those distances with how much the light from a supernova has shifted, they could calculate how the expansion rate has changed over time. For example, light that began its journey across space from a source 10 billion years ago would have a red shift markedly more pronounced than light that was emitted from a source just 1 billion years ago.

When Hubble made his measurements in 1929, the farthest red-shifted galaxies were roughly 6 million light years away. If expansion was now slowing under the influence of gravity, supernovas in those distant galaxies should appear brighter and closer than their red shifts would suggest. Instead, just the opposite was true. At high red shifts, the most distant supernovas are dimmer than they would be if the universe were slowing down. The only plausible explanation for this is that instead of gradually slowing down, the expansion of the universe is speeding up.

Scientists have devised an ingenious explanation for why this might be happening. "The driving force behind cosmic acceleration appears to be the repulsive gravity of a weird form of energy associated with nothing," says Turner. This unknown substance is the dark energy. If dark matter gives rise to the gravity that holds the universe together, then dark energy is the counterforce pushing the universe apart. At this stage of the game, dark energy appears to be winning the cosmic tug-of-war.

Here's what scientists think has happened so far. Very early in the existence of the universe, dark matter dominated. Everything was closer together, so the density of dark matter was higher than that of the dark energy, and the gravitational pull of dark matter was stronger than that of dark energy. This led to the clumping that formed early galaxies. But as the universe continued to expand, the density, and hence the gravitational pull, of dark matter decreased until it was less than that of the dark energy. So instead of the expected slowdown in the expansion rate, the now dominant dark energy began pushing the universe apart at ever faster rates.

Where does this dark energy come from? Scientists don't really know. But it's a testament to Einstein's genius that even his blunders prove to be significant. His "fudge factor," the cosmological constant called lambda, implied the existence of a repulsive form of gravity. And the simplest example of repulsive gravity can be found in quantum mechanics. In quantum physics, even a vacuum is teeming with energy in the form of "virtual" particles that wink into and out of existence, flying apart and coming together in an intricate quantum dance. (It appears that Spinoza was right in saying that nature truly abhors a vacuum.) This roiling sea of virtual particles could give rise to dark energy, giving the universe a little extra push so that it can continue accelerating.

The problem with this picture is that the numbers don't add up. The quantum vacuum contains too much energy: roughly 10^{120} times too much. So the universe should be accelerating much faster than it is. An alternative theory proposes that the universe may be filled with an even more exotic, fluctuating form of dark energy: "quintessence." Yet all the observations to date indicate that the dark energy is constant, not fluctuating. Scientists must therefore consider even more possibilities.

For example, the dark energy could be the result of the influence of unseen extra dimensions predicted by string theory. Alternatively, in 2004, scientists at the University of Washington proposed that the dark energy could be due to neutrinos—the lightest particles of matter—interacting with hypothetical particles called "accelerons." Some scientists have theorized that dark matter and dark energy emanate from the same source, although they don't know what that source might be. Yet it's just as likely that there is no connection, and the two are very different things. Or perhaps there is no such thing as dark energy, and general relativity is not an accurate description of gravity after all.

There's one thing scientists can agree on: thanks to cosmic acceleration, the shape of the universe will no longer determine its ultimate fate. Instead, its fate rests on whether the dark energy is constant or changing. "Until we better understand cosmic acceleration and the nature of the dark energy, we cannot hope to understand the destiny of the universe," says Turner. If the dark energy remains constant, the acceleration will continue indefinitely, and matter will grow farther and farther apart. Within 100 billion years, we will be able to see only a few hundred galaxies, compared with the hundreds of billions we can see today. If the quintessence model is correct, the fate of the cosmos depends on whether the dark energy decreases or increases over time. If the dark energy decreases, expansion could slow enough for the universe to recollapse in a big crunch. And if the dark energy increases its expansion rate over time, it could rip apart every galaxy, star, and atom in the universe within 100 billion years. Scientists call this the "big rip." Thus far, the dark energy appears to be constant.

The most surprising aspect of the "shadow particles" discovered by Pullman's Mary Malone is that they seem to be conscious, flocking to human thought like birds, and clustering around any objects that result from human creativity and workmanship. They are literally particles of consciousness, born from human thoughts and feelings. Dark energy isn't likely to be associated with consciousness, but that doesn't mean the universe lacks a memory. Some remnant of every kind of particle that has ever existed is still there—even though the particles themselves have long since disappeared. The evidence can be found in the equations physicists use to describe the various sub-atomic particles and fundamental forces that govern our universe. "The universe today contains only the stable relics and leftovers of the big bang," says Turner. "The unstable particles have decayed away with time, but the structure of space 'remembers' all the particles and forces we can no longer see around us." The equations never forget.

36

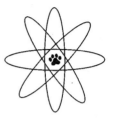

The Case of the
Missing Neutrinos

February 2001
Solution to solar neutrino problem

One gusty autumn evening—in a short story by Edgar Allan Poe—
a very perplexed prefect of the Paris police arrives at the home of
C. Auguste Dupin, a poet and man of letters with a penchant for solving puzzles. The prefect explains that a compromising letter has been
stolen from the royal apartments, and the honor of a woman of very
high standing is at stake. But Poe's story is far from a typical whodunit.
The perpetrator is already known: a scheming minister identified
simply as D—, who is using the letter to blackmail the woman in question. The police have been trying for three months to regain possession
of the letter and avoid an overt scandal. Having been given a detailed
description of the missive's outward appearance, they have searched
the minister's residence from top to bottom, and even waylaid him
twice in the hope of finding it concealed on his person—all to no avail.
Asked for his opinion, Dupin cryptically observes, "Perhaps the
mystery is a little *too* plain . . . a little *too* self-evident."

One month later, the prefect returns, now desperate to find the elusive epistle, and offers a reward of 50,000 francs to anyone who can
help him. In that case, Dupin says, he will hand the prefect the letter in
exchange for a check. And with panache worthy of Sherlock Holmes,
Dupin does present him with the letter, which in the interim he has
sneakily recovered from the minister's residence. "The Purloined
Letter" was written more than 100 years ago, yet it is an apt analogy

for a twentieth-century mystery that puzzled physicists for almost 35 years: the case of the missing solar neutrinos.

Neutrinos are tiny subatomic particles that travel very near the speed of light. John Updike's poem "Cosmic Gall" (1959) pays tribute to the two most defining features of neutrinos: they have no charge and a very low mass; until quite recently, physicists believed that they had no mass. Although they are the most abundant subatomic particle in the universe, they are extremely difficult to detect. That's because they very rarely interact with any type of matter. We are being bombarded by neutrinos, yet they pass right through us without our even noticing. Only one out of every trillion solar neutrinos would collide with an atom on its journey through the Earth; and a wall of ordinary matter more than 100 light-years thick would be needed to stop a beam of neutrinos. Isaac Asimov called them "ghost particles."

Why do scientists care so much about a ghostly particle so insubstantial that it almost isn't even there? For starters, neutrinos could provide clues about the nature of the hypothetical "dark matter" that scientists believe holds the universe together, as well as the "dark energy" they believe is causing the expansion of the universe to accelerate. And neutrinos are critical to understanding how the sun and other stars shine. Most of the neutrinos that reach Earth are produced in the deepest parts of the sun, and because they are so light, they can escape unimpeded (and thus largely unchanged) by collisions with other particles of matter. They can thus provide valuable clues to scientists about the inner workings of the solar core.

What do these spectral subatomic particles have to do with nineteenth-century literature? In Poe's story, the police know that the letter has been stolen. They know who stole it, and that the culprit still has it in his possession. They just can't find it. Physicists faced a similar predicament in the late 1960s. They knew neutrinos existed. They knew how many should be emitted by the sun, and thus how many should be detected on earth. But they were able to find only a fraction of the expected number.

Our scientific mystery begins in 1930, when the physicist Wolfgang Pauli proposed the existence of an unseen particle. Enrico Fermi later called it a "neutrino" and was the first to describe how neutrinos might interact with other particles, through something called the weak nuclear force. The weak force allows certain types of subatomic particles to exchange energy, mass, and charge—in essence, to change into each other. Fermi theorized that when a neutron inside an atom decays, it produces a proton, an electron, and a neutrino. This occurs 200^{36} times every second in the core of stars like our sun, as hydrogen is converted into heavier elements, beginning with helium. The process releases huge amounts of energy (in the form of heat, or sunlight), and neutrinos are a by-product of those reactions. Trillions of them are produced by the sun every day. Clyde Cowan and Frederick Reines observed the first neutrinos in 1956, using a nuclear reactor in South Carolina to generate the necessary fusion reactions.

Scientists wanted to detect solar neutrinos because doing so would confirm that the source of the sun's energy is a chain of thermonuclear reactions (fusion) at its core. Yet the feat was easier said than done. Scientists can sense the presence of neutrinos only if they can get neutrinos to interact with other subatomic particles of matter by means of the weak nuclear force. The problem is that the weak force is very weak indeed, and it kicks in only if neutrinos are so close as to be practically touching the atomic nuclei—like the ghosts of the departed in Thornton Wilder's play *Our Town*, who can be sensed only when the townspeople draw near to the graveyard. If a collision occurred, it would cause the atom to change into a different chemical element—an easily measurable effect.

That's how scientists conceived of building large neutrino observatories: enormous tanks of liquid buried deep inside the earth to avoid interference from other particles in the atmosphere, such as cosmic rays. The concept was that if enough neutrinos emitted by the sun flow through a large enough tank of fluid, occasionally one will strike an atom in the fluid, initiating the telltale decay process. It was a bit like trying to find one particular grain of sand in the midst of the Sahara Desert, but the physicist Ray Davis, Jr., thought it was worth a try. He built the first neutrino observatory in 1967 in an old mine in South Dakota called Homestake. He filled a giant tank with 600 tons of dry-cleaning fluid, which is primarily composed of chlorine. The rare instance of a neutrino colliding with a chlorine atom would create an argon atom that could then be detected as it decayed and emitted an electron.

The experiment worked: Davis was the first to detect neutrinos of solar origin. There was just one problem. Solar theorists had calculated that roughly 30 million neutrinos should pass through every cubic inch of earth every second, and predicted that therefore, the Homestake facility should be detecting one neutrino encounter per day. But Davis found far fewer than expected: one neutrino encounter only every three days. Roughly two-thirds of the expected solar neutrinos were missing. Working from the premise that bigger is better, experimentalists expanded their search. Even larger neutrino observatories were built, each using different detection methods, but there was still no trace of the missing neutrinos.

For the next three decades, scientists battled over why this was so in a back-and-forth game of finger-pointing. Solar theorists blamed the experimentalists, accusing them of poorly constructed experiments. Experimentalists countered by insisting that the theorists hadn't calculated the temperature and pressure of the sun's interior correctly. They weren't finding more neutrinos because there weren't any more neutrinos to be found. Both camps, it turned out, were incorrect; there was an unknown factor.

In Poe's story, the police followed procedure to perfection, but they assumed that any man who has stolen something will go to great pains to conceal it. Dupin reasoned that D—was both a poet and a mathematician, and as such combined logic with creativity. ("As mere mathematician, he could not have reasoned at all," Dupin snipes, clearly biased toward the poetic profession.) And he concluded that the minister would opt for simplicity, concealing the letter by hiding it in plain sight. Similarly, the neutrino mystery wouldn't be solved until scientists at the Sudbury Neutrino Observatory (SNO) in Ontario, Canada, proved that the missing neutrinos had been right under their noses all along. The experimental procedures had been flawless. Scientists were just looking for the wrong type of neutrino.

Neutrinos come in different flavors, each with a different mass: we'll call them orange, cherry, and grape. Electron (orange) neutrinos are produced during the nuclear fusion of the sun. Muon (cherry) neutrinos and tau (grape) neutrinos are produced only by decay processes that require higher energy; the latter, for instance, are produced only during supernova explosions. The sun produces only orange-flavor neutrinos, so all prior experiments were set up to detect that one flavor, and to ignore the cherry and grape neutrinos. The SNO detector, however, was designed to look for the other two flavors as well.

For a detector fluid, the SNO facility uses heavy water, in which two hydrogen atoms in each water molecule are replaced with a hydrogen isotope called deuterium. When a neutrino collides with a deuterium atom in heavy water—this occurs roughly 10 times a day—the atom is ripped apart. There are two different ways the atom can split. If it is an orange neutrino, the reaction produces two protons and an electron, generating a flash of light. By measuring those flashes, scientists can measure the number of orange neutrinos. To find the cherry and grape neutrinos, SNO scientists simply added salt to the heavy water. Salt is made of sodium and chlorine atoms. If a cherry or grape neutrino collides with a deuterium atom, the atom splits into a proton and neutron. The chlorine atoms in the salt will absorb the neutron and emit gamma rays—the sort of high-energy radiation that is very easy to detect. When the researchers added all three flavors together, the total number detected agreed with theoretical predictions for solar neutrinos. Yet scientists knew that the sun emits only orange-flavor neutrinos. Clearly, something was happening to some of the neutrinos as they journeyed through space to Earth.

In 2002, the SNO team confirmed that it had found the "missing neutrinos." They weren't missing after all: they were just in disguise. When Poe's amateur gumshoe, Dupin, visits the guilty minister, he notices a card rack dangling by a dirty blue ribbon from the mantel. It holds a soiled, crumpled letter that bears almost no similarity to the one that was stolen. Yet closer inspection reveals that it is the purloined letter, turned inside out to alter its appearance and disguise its true nature. That's exactly what solar neutrinos do as they travel from the sun to Earth: they "change clothes," or flavors, in transit to alter their appearance and thus escape detection. Physicists have called this neat little trick "neutrino oscillation." Different-flavor neutrinos can be compared to piano strings, which are tuned to specific notes: let's say G, B, and C. Scientists assumed that if a neutrino was born as a G, it would always be a G. But it turns out that neutrinos can "de-tune" over time, just like the strings on a piano. So a G can gradually become a B or a C, and so on.

As often occurs in science, the mystery is only half solved—or rather, solving the case has raised new questions. We know that neutrinos can change flavors on the fly, but we don't know why or how they are able to do so. And in order to change flavors, neutrinos must have some tiny, barely perceptible mass, which doesn't agree with the working standard model of particle physics. There are now as many competing theories about neutrino oscillations as there were to explain the missing neutrinos.

Today, astronomers hope to use neutrino detection to help them understand what is happening during the explosions of dying stars, called supernovas. They can't detect most of the light from supernovas in the Milky Way galaxy, because the light is obscured by galactic dust. Neutrinos pass right through the dust and thus arrive on Earth with the "information" they carry fully intact.

On February 23, 1987, the Canadian astronomer Ian Shelton was engaged in what he thought was routine work at Las Campanas Observatory in Chile: taking a telescopic photo of a small galaxy 167,000 light-years from Earth, called the Large Magellanic Cloud. But when he developed the photographic plate, he noticed an extremely bright star that had not been there before. It wasn't a new star but an aging massive star that had blown apart in a supernova explosion. Astronomers named it 1987A. At the same time, the Super-Kamiokande neutrino detector in Japan picked up a few extra neutrinos.

A supernova emits 1000 times more neutrinos than the sun will produce in its entire 10-billion-year lifetime. Yet the Super-Kamiokande detector picked up only 19. It was enough to confirm predictions that a supernova would produce a sharp pulse of neutrinos. But to understand how supernovas really work will require the detection of thousands of neutrinos from a single supernova. Scientists now must unravel a mystery that is far from plain and self-evident—a challenge that even Poe's Dupin might find worthy of his rarefied intellect.

37

Icarus
Descending

September 2002
Schön found guilty of scientific misconduct

The plot of Dorothy Sayers's mystery novel *Gaudy Night* (1936) rests not on murder but on a case of falsification of data. Near the completion of his magnum opus, a historian discovers a document containing irrefutable evidence against the hypothesis he has spent years painstakingly constructing. Too good a historian to destroy the document, he steals it instead. But his theft is discovered, he loses his post and professional reputation, and he ultimately commits suicide. The explanation offered is that the historian had been working on the project for so long that he became enamored of his own theory and couldn't bear to give it up. "That's the mark of an unsound scholar, I'm afraid," an Oxford don sadly remarks.

A similar incident occurs in C. P. Snow's novel *The Search*. A scientist who makes a careless error in a paper, and doesn't correct it, loses a prestigious academic position. Sayers's fictional detective, Lord Peter Wimsey, draws a parallel between Snow's novel and *Gaudy Night's* ill-fated historian. He argues that while this might seem to be a very harsh punishment for a careless error, it is necessary to protect the integrity of science. "The only ethical principle which has made science possible is that the truth shall be told all the time," he quotes a senior scientist in Snow's novel as saying. "If we do not penalize false statements made in error, we open up the way for false statements by intention. And a false statement of fact, made deliberately, is the most serious crime a scientist can commit."

As long as there has been science, there have been hoaxes and outright fraud. Modern scientists define fraud as making up data, changing data or results to misrepresent experiments, or plagiarizing the work of other scientists. It is because of the possibility of fraud that scientists must be able to reproduce any noteworthy experimental results in order to establish scientific fact. For example, a few years after the discovery of X rays in 1895, a French physicist, René Blondlot, announced that he had discovered a new kind of radiation called N rays, which could be detected only with a special machine of his own devising. An American colleague publicly debunked his claims by secretly removing the inner nuts and bolts of the machine. Blondlot repeated the demonstration, and, to the American's amusement, still claimed to see N rays, unaware that his "detector" no longer worked.

But fraud isn't always a case of chicanery. Motives can be mixed, hard to pin down, and as inscrutable as the individuals themselves. Sometimes, as in Snow's novel, fraud results from carelessness or error in the rush to be first to publish groundbreaking results in the intensely competitive research environment. Sometimes it results from naked ambition and greed, or the desire to "put one over" on colleagues or the public at large. For instance, in 1726 spiteful colleagues of a German naturalist, named Johann Berginger, planted fake fossils of birds, frogs, and spiders outside his hometown of Würzburg. They also planted fake stones bearing the Hebrew letters for Jehovah. Before the stones were unmasked as a hoax, Berginger, a devout Christian, published a book based on his findings, claiming the stones as evidence of naturally occurring examples of the relationship between God and nature. The fake fossils became known as the *Gensteine* ("lying stones").

Sometimes, as with Sayers's fictional historian, the perpetrators of a fraud cling to pet theories and imagined discoveries despite all evidence to the contrary. An amusing "letter" has been floating around the Internet for years, purportedly from the curator of antiquities at the Smithsonian to an amateur archaeologist in Vermont named Scott Williams. Williams digs up random objects in his backyard and sends these "specimens" to the Smithsonian, insisting that they are genuine archaeological finds. His prodigious imagination transforms the chewed-up head of a Barbie doll into a 2-million-year-old hominid skull, and a rusty crescent wrench into a femur from a juvenile *Tyrannosaurus rex*. He devises bizarre hypotheses in support of his conclusions—attributing the bite marks on Barbie's head to ravenous man-eating Pliocene clams—and ignores evidence to the contrary (the bite marks are clearly from a dog). Actually, however, there is no such deluded individual; the letter is a "netmyth," the Internet equivalent of an urban legend. But it's an excellent example of the human propensity to twist facts to fit a pet theory.

No scientific discipline is immune to fraud. Biology was beset by numerous cases in the 1980s, although the most celebrated turned out to be a false accusation. In 1985, a junior colleague accused Teresa Imanishi-Kari—a Brazilian-Japanese biologist then at MIT—of falsifying data in a published paper. The incident provoked a media feeding frenzy, including a cover story in *Time* magazine. It also captured the attention of Congress, most notably John Dingell, a Democrat from Michigan who held hearings on the subject, and even called in the U.S. Secret Service to examine notebooks and data tapes from the instruments used in the experiment. Imanishi-Kari's coauthor, the Nobel-prize-winning biologist David Baltimore, staunchly defended her. His loyalty cost him several friendships and an appointment as president of Rockefeller University, while Imanishi-Kari's funding grants were terminated and her tenure was postponed. She became a symbol for everything that was wrong with science. Yet several other labs later confirmed her findings. Her exoneration barely warranted passing mention in the press—and certainly not the cover of *Time*.

While biologists struggled to rebuild their credibility, physicists complacently assured themselves that such things could never happen in their field. They were in for a rude awakening. In 2002, physics was hit with back-to-back cases of alleged fraud. And both allegations turned out to be true. The trouble started in 1999 with an experiment at Lawrence Berkeley National Laboratory in which researchers bombarded a lead target with particles of krypton at very high speeds and studied the decay of the smashed bits of the atoms. They claimed to have produced element 118, the heaviest atom ever found. But subsequent experiments failed to reproduce the results, and a reanalysis of the original data did not detect the telltale atomic decay patterns that had supposedly indicated the existence of element 118. Within three years, the discovery was withdrawn and the laboratory fired the head physicist Victor Ninov for falsifying the data that "proved" the claim.

The physics community was still reeling from that debacle when suspicions arose about the work of Lucent Technologies/Bell Labs' German-born wunderkind, Jan Hendrik Schön. In 2001, Schön attached gold electrodes to a pentacene molecule (a material that doesn't normally conduct electricity) and covered them with a layer of aluminum oxide. This turned the molecule into a tiny conducting transistor, which he used to build simple solar cells and lasers. Then he discovered that so-called buckyballs—carbon molecules made out of 60 carbon atoms linked into the shape of a soccer ball—could become superconductors at far higher temperatures than had previously been believed. A superconductor is a material that can conduct electrical current at high speeds with very little friction.

Some have likened Schön to a modern-day alchemist, because he magically caused electricity to boldly conduct in materials where it had never conducted before. Barely five years after finishing graduate school, Schön was being mentioned as a contender for the Nobel Prize in physics, and he'd accepted a position as director of one of the Max Planck Institutes in Germany. But there were also whispers that his results were more science fiction than fact.

First, no one else could duplicate his results. Second, the physicist Paul McEuen at Cornell and his colleague Lydia Sohn at Princeton noticed that diagrams in two separate papers published in the journals *Nature* and *Science* supposedly described the electrical behavior of two different materials. Yet on closer analysis, it was clear to any trained scientist that the diagrams were either identical or based on identical data records. Schön had used the same graphs to illustrate the outcomes of two different experiments. McEuen and Sohn sounded the alarm. Schön insisted that the duplication was merely a clerical error, offering substitute diagrams and declaring, "I haven't done anything wrong." But subsequent investigation turned up several more similar instances, and Bell Labs launched a full-scale investigation.

Hindsight is 20/20, and in retrospect, the faked data aren't all that convincing. "The data were too clean," one researcher told *Salon*. "They were what you would expect theoretically, not experimentally." A panel of scientific experts examined more than two dozen of Schön's published papers and found "compelling evidence" of manipulation and misrepresentation of data on 16 separate occasions. Schön had substituted entire figures from other papers, removed data points that disagreed with his theoretical predictions, and even used mathematical functions (theory) in place of actual observed data. Furthermore, he had done this "intentionally and recklessly and without the knowledge of any of his coauthors," and in so doing, showed "a reckless disregard for the sanctity of data in the value system of science."

Schön's supervisor and coauthor on many of the questionable papers, Bertram Batlogg, also came under fire for having lent his name to papers when he had never witnessed any of the experiments. Senior researchers routinely sign off on the lab notes of their younger colleagues, or ask for raw data files on experiments. Batlogg didn't, yet his standing in the physics community gave Schön's results a perceived stamp of approval. Batlogg admitted as much in an apologetic e-mail to colleagues, adding that he had "placed too much trust" in his collaborator.

The commission that investigated the case of element 118 made similar criticisms of Victor Ninov's coauthors. "Given the importance of the result, it was incredible that, prior to publication, no one had looked at the raw data for the particular events claimed to make sure there had been no errors," said George Trilling, a prominent physicist who chaired the commission. "Extraordinary results demand extraordinary supporting evidence, and the burden of proof for an unexpected or major discovery is much greater than for a routine measurement."

Not that asking for the raw data files would have done any good in Schön's case. He couldn't produce the data when the investigating panel requested them, and he claimed that he had deleted the files because his computer lacked sufficient memory—a ludicrous excuse in an age when computer memory is increasing exponentially. And his magical single-molecule transistors had been destroyed. The very act of deleting his data records constituted scientific misconduct. To date, neither Schön nor anyone else has been able to replicate the results.

The commission released its report in September 2002. Once Schön's guilt had been established, consequences fell hard and fast.

Bell Labs summarily dismissed Schön from its staff, the Max Planck Institute withdrew his directorial appointment, he was stripped of several prestigious prizes, several journals retracted questionable papers, and he lost his PhD. He returned in disgrace to Germany and quickly faded into obscurity. But the repercussions went beyond Schön himself. More than 100 laboratory groups around the world had been working on duplicating his results. Graduate students had based their PhD theses on those experiments, and postdoctoral fellows had hoped to build on Schön's work to make their way into prestigious positions. Also, the incident undermined public confidence in science.

If the scientific method works so well, why did it take four years from the first time Schön published falsified data for his dishonesty to come to light? Part of the answer is that there are no set rules about how fast scientific knowledge should advance. It takes time and money to design and conduct good experiments, and they don't always work on the first attempt. Blondlot's N rays were debunked immediately, but it was more than 100 years after Copernicus before scientists accepted that the sun and not the Earth was the center of the solar system. It took 10 years before Imanishi-Kari was vindicated; she was tried and convicted in the court of public opinion before anyone had even attempted to duplicate her results. A four-year interval between the perpetration of a fraud and its unmasking doesn't seem all that unreasonable in comparison.

The other factor is that physics is not the abstract realm of absolutes that many people perceive it to be. There are as many open questions and elusive mysteries as there are hard established facts. Science is very much a human endeavor, and scientists are as fallible as the rest of us. The sheer number of papers Schön was producing should have aroused suspicion. Most scientists can manage only about three significant articles per year. Schön pumped out more than 90 papers in three years. "I find it hard to read that many papers, much less write them," the physicist Art Ramirez told *New Scientist*; and the Nobel laureate Philip Anderson admitted to *Salon* that in Schön's case the physics community had been guilty of extreme gullibility: "We should have all been suspicious of the data almost immediately." Like Fox Mulder in *The X-Files*, physicists wanted to believe. That desire not only blinded them to the possibility of fraud but also helped perpetuate it.

Schön's story is an all too human tragedy. At the height of his brief career, he was often compared to the legendary king Midas, since every experiment he touched seemed to turn into scientific gold. In the wake of the scandal, he seems more like Icarus, who fastened fabricated wings to his shoulders with wax so that he could fly. Icarus flew too close to the sun, which melted the wax, and he plunged to his death. Had Schön's faked findings been less revolutionary, they would not have attracted such close scrutiny, and he might never have been caught. Why did he do it? Naked ambition and the pressure to produce results in a highly competitive environment seem to be inadequate motives. Schön was too good a scientist not to realize that sooner or later, the inability of others to duplicate his results would unmask his deceit.

"If the results are fraudulent, Schön would have to have some kind of psychological problem," one scientist opined anonymously when the accusations first came to light. Yet it's quite possible that Schön truly believed his work would eventually be vindicated and his impatient tweaking of the data would never be discovered. He indicated as much in his last public statement, in which he professed not only his innocence but his unshaken belief that it was only a matter of time before scientists succeeded in building a single-molecule transistor.

What if Schön was telling the truth, and the duplicated graphs really were just a clerical error? Did he still deserve such severe punishment? In *Gaudy Night*, the fraudulent historian is at the center of a discussion between Wimsey and several Oxford dons about the often harsh nature of adhering strictly to principle. "The first thing a principle does—if it is really a principle—is kill somebody," Wimsey says. This means, not that the principle is wrong, but just that it comes with a price.

38

String
Section

October 2003
Nova special on string theory

In *Gut Symmetries*, the British novelist Jeanette Winterson explores the complex relationships between a middle-aged string theorist named Jove; his wife, Stella; and his mistress, Alice, a budding young physicist who also becomes Stella's lover when the two eventually meet. "Jove only works on superstrings because it reminds him of spaghettis," his Italian mother jokes. In truth, both Jove and Alice are drawn to the field because it skirts the farthest, most rarefied reaches of science, where physics and mathematics impinge on the intangible realm of metaphysics and metaphor.

According to string theory, the universe is made of tiny string-like objects that vibrate and wriggle in different ways. How the strings wriggle determines what kind of elementary particles are formed, and generates the physical forces we observe around us, in much the same way that vibrating fields of electricity and magnetism give rise to the entire spectrum of light. String theory is the best candidate thus far for a so-called grand unified theory (GUT), master equations describing the workings of the entire universe at every size scale. Scientists have been searching for a GUT ever since Albert Einstein embarked on a fruitless 30-year quest to merge all the known physical forces into one, an approach that dates back to the unification of electricity and magnetism in the nineteenth century. Einstein successfully unified space and time, energy and mass, and gravity and acceleration, but his ultimate goal eluded him. A true GUT would be like having a skeleton key capable of unlocking the most basic secrets of the universe.

Since the 1920s, physics has been a field divided, at least in terms of its theoretical underpinnings. There are two rival "gangs" in physics, like the Sharks and the Jets in *West Side Story*. First there are the Relativistic Rebels, who use general relativity to study objects and physical systems on the macroscale, from cats to cars to the cosmos. Then there are the Kings of Quanta, who apply the principles of quantum mechanics to the behavior of objects and systems at the atomic level. Neither side fares very well when it ventures onto the other's turf: relativity breaks down at the atomic level, and quantum phenomena don't materialize on the macroscale. (When was the last time you saw anyone walk through a wall? But at the quantum level, atoms "tunnel" their way through seemingly impenetrable energy barriers all the time.) That's the conundrum currently facing physics. Physicists in search of a grand unified theory have the unenviable task of trying to meld relativity and quantum mechanics into one single theory to describe the universe at all size scales. This task is as difficult as mixing oil and water, or getting the quarrelsome Sharks and Jets to merge into one peace-loving supergang.

There are four fundamental forces in modern physics: gravity, which is governed by general relativity; and electromagnetism, the strong nuclear force, and the weak nuclear force, all three governed by the laws of quantum mechanics. The four forces have different strengths and apply to different size scales. Gravity is the means by which planets orbit stars, whereas electromagnetism causes electrons to orbit atomic nuclei. Within a nucleus, the strong and weak forces come into play. The strong force holds protons and neutrons together to form a nucleus, and when that force is overcome to split the atom, vast amounts of energy are released. This is what happens in nuclear fission and fusion. The weak force controls interactions between atomic nuclei and neutrinos, the tiny particles created when neutrons become protons, or vice versa. The weak force kicks in only if neutrinos are so close as to be practically touching the atomic nuclei. When this happens the atom is turned into a different chemical element.

Meshing such disparate forces together is not an easy task. The fires of passion bind Winterson's three lovers, and a similarly intense heat is required to merge physical forces. At sufficiently high temperatures, electromagnetism and the weak force merge and become the electroweak force, just as ice cubes in a glass of water will melt and merge with surrounding liquid on a hot summer day. Scientists believe that if temperatures are increased even further—to levels that existed only in the first fractions of a second after the big bang—the electroweak and strong force will also merge and form one universal force. This is the so-called "standard model" of particle physics.

The sticking point is gravity. Jove puckishly dubs the standard model "the flying tarpaulin," because it is "big, ugly, useful, covers what you want and ignores gravity." But gravity is always part of the equation. Alice knows this, yet she embarks on not one, but two, torrid affairs in a vain attempt to defeat it, and inevitably comes crashing to earth. "It was a volatile experiment, soon snared by the ordinariness we set out to resist" she admits. In physics, as in love, gravity is the ultimate spoiler. The standard model achieves its harmonious symmetry only by turning a blind eye to the inescapable reality of gravity. The fact that gravity travels at the speed of light indicates an underlying connection between the two. But the two forces are vastly different in strength. We think of gravity as a powerful force—after all, it holds the planets in their orbits. Yet we overcome it in all kinds of little ways every day. This is because, at the atomic level, gravity appears to be feeble, easily overpowered by the violent fluctuations that typify the quantum realm.

Still, both relativity and quantum mechanics function perfectly well at their assigned size scales, so why try to reconcile the two? Physicists are reductionists at heart; like the medieval philosopher William of Ockham, they hate to multiply entities beyond necessity. So having two competing sets of rules for the universe strikes them as untidy. Also, the situation can give rise to unwieldy anomalies. Sometimes the wife and the mistress meet. There are some exotic cases where both sets of rules can apply.

Consider the phenomenon of black holes. A black hole occurs when a star implodes and its entire mass is concentrated into an ultra-small spherical region. This warps the surrounding space-time so profoundly that anything, including light, that gets too close won't be able to escape the gravitational pull. This point is known as the event horizon. Objects that cross the event horizon and fall into the black hole would be subject to an ever-increasing gravitational strain as they fall toward the center.

The center of a black hole is known as its singularity. It is both infinitely dense, in terms of mass, and infinitely small. Here's the puzzle: Should scientists apply the rules of general relativity to study black holes because a star is so massive; or should they use the principles of quantum mechanics because the singularity is so tiny? By their nature, black holes are a paradox: the center of a black hole is both massively heavy and exceedingly tiny. This has important ramifications for the big bang theory of the origin of the universe. Scientists believe that at its birth, the observable universe resembled just such a singularity: infinitely dense and tiny. A master set of equations that could describe the universe at all size scales would enable physicists to resolve this strange anomaly.

This is where string theory comes into play. It dates back to 1968, when a young postdoc, Gabriele Veneziano, unearthed a forgotten 200-year-old equation by the Swiss mathematician Leonhard Euler, which had been dismissed by Euler's contemporaries as a bizarre curiosity. Veneziano realized that the equation was an accurate description of the strong force—completely unknown in Euler's time. A fellow theorist, Leonard Susskind, heard of Veneziano's discovery and did his own calculations. He found that Euler's equation also described vibrating elastic particles very accurately, and reenvisioned elementary particles as strings. Such strings could stretch, contract, and wriggle, like violin or cello strings. And if the pieces of string were minuscule enough, they would still appear to be the solid, fixed, pointlike particles experimentally produced and observed in atom smashers around the world.

Susskind thought he'd made a momentous discovery, but his first paper on the subject was summarily rejected by his scientific peers. The notion was just too odd. He and a few other theorists persisted, but string theory was an academic backwater for the next 15 years, mostly because many of its predictions conflicted with experimental observations. It was also plagued by mathematical singularities; more often than not, an equation would yield a meaningless answer: infinity. That's generally an indication to physicists that an equation is wrong, since there are no infinities in the physical world. Then, in the summer of 1984, two mathematicians succeeded in removing the singularities, proving that string theory was sufficiently broad to encompass all four fundamental forces and all of matter. This made it a bona fide GUT.

String theory suddenly became a legitimate field; in fact, it exploded, with more than 1000 research papers on the subject published over the next two years. But it also became a victim of its own success. So many people were working in the area that by the late 1980s, there were five different versions of string theory, with many common elements but vastly different mathematical details. Which was the right one? In 1995, a theoretical physicist, Ed Witten, caused a sensation at a scientific conference on string theory when he announced that he had found a way to boil down the five competing theories into one. According to Witten, there weren't five different theories, just five different ways of looking at the same thing, like gazing at reflections in a mirror. Four of the competing theories were actually just mirror images. Witten calls his version the M theory (no one seems to know what the M stands for; and if Witten knows, he isn't telling).

According to string theory, any particle, when sufficiently magnified, would be seen not as a solid fixed point but as a tiny vibrating string: a one-dimensional filament akin to an infinitely thin rubber band. Formed in the earliest seconds of the universe, these strings are so small that they appear pointlike. All matter, and all forces, are composed of these vibrations; that is what makes string theory such a good candidate for a GUT. Strings line the entire universe, and even though they are thinner than an atom, they can still generate an enormous amount of gravitational pull on any objects that pass near them. String theory also explains why elementary particles have certain masses, force charges, and relative strengths of the forces—all of which are known from experimental observations, but with little to no theoretical understanding as to how or why those specific properties exist.

There's a catch, of course. In order to bring all these disparate pieces together into a coherent whole, string theory requires extra dimensions of space. We're accustomed to the dimensions we can sense: the three dimensions of physical space, and the fourth dimension of time.

String theory calls for 11 separate dimensions: the four we know and love, plus another seven dimensions, curled up at every point in space-time, like the circular loops of thread in a carpet's pile. These extra dimensions give rise to the fundamental constants that govern the known universe: the specific masses and charges of elementary particles, for instance, and the four forces. In his best-selling book *The Elegant Universe*, the string theorist Brian Greene draws a musical analogy, asserting that the extra dimensions "keep the cosmic symphony of strings in tune." Like a stage manager backstage, they are never seen, but their contributions are vital to the performance.

If extra dimensions exist, why can't we sense them? The dimensions are very, very small, beyond detection even by today's cutting-edge instruments, which are capable of measuring objects as small as 1 quadrillionth of a billionth of a meter. We could pass through the extra dimensions hundreds of times every day and never be aware of them. They are crumpled up like wads of paper, not into just any random shapes, but into a class of exotic six-dimensional geometric shapes known as Calabi-Yaus.

**The 11 dimensions of space-time
in string theory**

As tiny as these dimensions are, strings are even smaller. A string moves around through 11 separate dimensions of space-time, oscillating as it travels, and the geometric form of the extra dimensions helps determine the resonant patterns of vibration. Those patterns are observed in turn by scientists as the masses and charges of the elementary particles that make up matter. The existence of extra dimensions might also explain why gravity appears to be so weak. According to Greene, gravity isn't weak. We just can't feel the full impact of its strength in our limited dimensional reference frame, because it dissipates into these extra dimensions. Gravity might behave like sound waves, able to escape the confines of its source into other-dimensional "rooms," thus diluting its strength.

In Winterson's novel, Alice muses that if string theory is correct, all matter is therefore notional, made not so much out of solid particles as out of intangible vibrations. "There is no basic building block, no firm stable first principle on which to pile the rest," she says. String theorists would argue that finding the firm stable first principle is actually the aim of their field. The real issue is that science currently lacks the means to directly observe such strings; their existence can only be inferred. This skirts dangerously close to taking string theory solely on faith in the underlying mathematics. If a theory can't be tested, it is not so much a science as a philosophy, at least in the minds of physicists—no matter how impressively complex and elegant the math. Even the noted string theorist James Gates has said, "If string theory cannot provide a testable prediction, then nobody should believe it." Physics, which began as natural philosophy, now appears to be coming full circle. It may turn out that ancient Greek and Renaissance philosophers were right all along when they mused upon the celestial "music of the spheres": the universe, in essence, may indeed be one cosmic string section.

Bibliography

Addams Family Values. Dir. Barry Sonnenfeld, 1993, Paramount.

Anderson, Bonnie S., and Judith P. Zinsser. *A History of Their Own: Women in Europe from Prehistory to the Present*, Vol. 2. New York: Harper and Row, 1988.

Anderson, David. *The Discovery of the Electron.* New York: Taylor and Francis, 1997.

Apollo 13. Dir. Ron Howard, 1995, Universal City Studios.

Atalay, Bulent. *Math and the Mona Lisa.* Washington, D.C.: Smithsonian Institution, 2004.

Auburn, David. *Proof.* New York: Faber and Faber, 2001.

Austin Powers: International Man of Mystery. Dir. Jay Roach, 1997, New Line Cinema.

Back to the Future. Dir. Robert Zemeckis, 1985, Universal Studios.

Baldwin, Neil. *Edison: Inventing the Century.* New York: Hyperion, 1995.

Ball, Philip. "New Model of Expanding Universe: There Are More Than Two Ways to Pump Up the Universe," *Nature*, February 2, 2001.

Batman Forever. Dir. Joel Schumacher, 1995, Warner Studios.

Baum, Rudy. "Drexler and Smalley Make the Case for and against 'Molecular Assemblers,'" *Chemical and Engineering News*, December 1, 2003.

Bennett, J.A., ed. *London's Leonardo: The Life and Work of Robert Hooke.* Oxford: Oxford University Press, May 2003.

Benyus, Janine. *Biomimicry: Invention Inspired by Nature.* New York: HarperCollins/Perennial, 2002.

Blade Runner. Dir. Ridley Scott, 1982, Ladd Company (through Warner Brothers).

Bloomfield, Louis. *How Things Work: The Physics of Everyday Life*, 2nd ed. New York: Wiley, 2001.

Bodanis, David. $E = mc^2$: *A Biography of the World's Most Famous Equation.* New York: Walker, 2000.

Bradbury, Ray. "I Sing the Body Electric." In *I Sing the Body Electric and Other Stories.* New York: HarperCollins/Perennial, 2001.

Bramly, Serge. *Leonardo: The Artist and the Man.* New York: Penguin, 1994.

Bringsjord, Selmer. "Chess Is Too Easy," *MIT Technology Review*, March–April 1998.

———. "A Contrarian Future for Minds and Machines," *Chronicle of Higher Education*, November 3, 2000.

———, and David Ferrucci. *Artificial Intelligence and Literary Creativity: Inside the Mind of Brutus, a Storytelling Machine*. Mahwah, N.J.: Lawrence Erlbaum, 2000.

Bromberg, Joan L. "The Birth of the Laser" *Physics Today*, October 1988.

Bromley, Allan. "The Evolution of Babbage's Calculating Engines," *Annals of the History of Computing 9*, 1987, 113–136.

Brown, Dan. *The Da Vinci Code*. New York: Doubleday, 2003.

Brumfield, Geoff. "Physicist Found Guilty of Misconduct," *Nature*, September 26, 2002.

Brunton, Michael. "Little Worries," *Time Europe*, May 4, 2003.

Brush, Stephen J. "Creationism versus Physical Sciences," *APS News*, November 2000.

Buffy the Vampire Slayer. "The Wish," Season 3. Dir. David Greenwalt, written by Marti Noxon. Air date: December 8, 1998. Twentieth Century Fox.

———. "Beer Bad," Season 4. Dir. David Solomon, written by Tracy Forbes. Air date: November 2, 1999. Twentieth Century Fox.

Burgess, Daniel. "A Hairy Worm Wears Photonic Crystals," *Photonics Technology*, March 2001.

Butterfly Effect, The. Dir. Eric Bress, 2004, New Line Cinema.

Čapek, Karel. *R.U.R.* Mineola, N.Y.: Dover Thrift Edition, 2001.

Carter, Thomas Francis. *The Invention of Printing in China and Its Spread Westward*. New York: The Ronald Press, 1925.

Cartnell, Robert. *The Incredible Scream Machine: A History of the Roller Coaster*. Bowling Green, Ohio: Bowling Green State University Popular Press, 1987.

Cassidy, David. *Uncertainty: The Life and Times of Werner Heisenberg*. New York: Freeman, 1992.

Cassuto, Leon. "Big Trouble in the World of Big Physics," *Salon*, September 16, 2002.

Chang, Kenneth. "A Sudden Host of Questions on Bell Labs Breakthroughs," *New York Times*, May 28, 2002.

———. "On Scientific Fakery and the Systems to Catch It," *New York Times*, October 15, 2002.

Chapman, Allan. "England's Leonardo: Robert Hooke and the Art of Experiment in Restoration England." *Proceedings of the Royal Institution 67*, 1996, 239–275.

Cheney, Margaret. *Tesla: Man Out of Time*. New York: Touchstone, 2001.

Clary, David. *Rocket Man: Robert H. Goddard and the Birth of the Space Age.* New York: Hyperion, 2003.

Crichton, Michael. *Jurassic Park.* New York: Ballantine, 1990.

——. *Prey.* New York: HarperCollins/Avon, 2002.

Cull, Selby. "Understanding the Ghost *Particle,*" *Journal of Young Investigators,* December 2003.

Dahl, Per. *Flash of the Cathode Rays: A History of J. J. Thomson's Electron.* London: Institute of Physics, 1997.

Devlin, Keith. "Cracking the Da Vinci Code," *Discover,* June 2004.

Dilson, Jesse. *The Abacus: A Pocket Computer.* New York: St. Martin's, 1968.

Doctor Strangelove, or How I Learned to Stop Worrying and Love the Bomb. Dir. Stanley Kubrick, 1964, Tri-Star Pictures.

Donne, John. "A Valediction: Forbidding Mourning," In *John Donne: The Complete English Poems.* New York: Penguin, 1983.

——. "A Valediction: Of Weeping," In *John Donne: The Complete English Poems.* New York: Penguin, 1983.

Eskin, Blake. "On the Roof: Pepsi Degeneration," *New Yorker,* March 29, 2004.

Feynman, Richard. *QED: The Strange Theory of Light and Matter.* Princeton, N.J.: Princeton University Press, 1985.

——. Six *Easy Pieces.* Cambridge, Mass.: Perseus, 1995.

——. *Six Not-So-Easy Pieces.* Cambridge, Mass.: Perseus, 1997.

Fournier, Marian. *The Fabric of Life: Microscopy in the Seventeenth Century.* Baltimore, Md.: Johns Hopkins University Press, 1996.

Foust, Jeff. "Solar Neutrino Problem Solved," *Spaceflight Now,* June 20, 2001.

Frankel, Felice. "Richard Feynman's Diagrams," *American Scientist,* September–October 2003.

Frayn, Michael. *Copenhagen.* New York: Random House/Anchor, 1998.

Freeman, Allyn, and Bob Goden. *Why Didn't I Think of That? Bizarre Origins of Ingenious Inventions We Couldn't Live Without.* New York: Wiley, 1997.

Gaiman, Neil. *American Gods.* New York: HarperCollins/Morrow, 2001.

——, and Terry Pratchett. *Good Omens.* New York: Workman, 1990.

Garcia, Kimberly. *William Roentgen and the Discovery of X-Rays.* Newark, Del.: Mitchell Lane, 2002.

Gibson, William, and Bruce Sterling. *The Difference Engine.* New York: Spectra, 1992.

Gilmore, Robert. *Alice in Quantumland.* New York: Springer-Verlag, 1995.

Gingerich, Owen. *The Eye of Heaven.* Woodbury, N.Y.: American Institute of Physics Press, 1993.

——. *The Book Nobody Read: Chasing the Revolutions of Nicolaus Copernicus.* New York: Walker, 2004.

Gleick, James. *Chaos: Making a New Science.* New York: Viking/Penguin, 1987.
——. *Genius: The Life and Science of Richard Feynman.* New York: Vintage, 1993.
——. *Isaac Newton.* New York: Pantheon, 2003.

Goldman, M. *The Demon in the Aether.* Edinburgh: Paul Harris, 1983.

Greene, Brian. *The Elegant Universe.* New York: Random House/Vintage, 2000.

Gribbin, John. *In Search of Schrödinger's Cat.* New York: Bantam, 1984.
——, and Mary Gribbin. *The Science of Philip Pullman's His Dark Materials.* London: Hodder Children's Books, 2003.

Hecht, Jeff. "Winning the Laser Patent War," *Laser Focus World,* December 1994.

Highfield, Roger. *The Science of Harry Potter.* New York: Viking/Penguin, 2003.
——. "Prince Charles Asks Scientists to Look into 'Grey Goo,'" *Daily Telegraph,* May 6, 2003.

Hockney, David. *Secret Knowledge: Rediscovering the Lost Techniques of the Old Masters.* New York: Penguin/Viking Studio, 2000.

Hopkins, Gerard Manley. "As Kingfishers Catch Fire." In *Poems and Prose of Gerard Manley Hopkins,* W. H. Gardner, ed. New York: Penguin, 1983.

Horgan, John. "Schrödinger's Cation: Physicists Prove That an Atom Can Be in Two Different Places at Once," *Scientific American,* June 17, 1996.

Hounshall, D. A. "Two Paths to the Telephone," *Scientific American,* 244, 1981, 156–163.

Howard, Toby. "The Spinach Machine," *Personal Computer World,* July 2000.

Hyman, Anthony. *Charles Babbage, Pioneer of the Computer.* Oxford: Oxford University Press, 1982.

Isaacson, Walter. *Benjamin Franklin: An American Life.* New York: Simon and Schuster, 2004.

It's a Wonderful Life. Dir. Frank Capra, 1947, Republic Studios.

Jardine, Lisa. *The Curious Life of Robert Hooke: The Man Who Measured London.* New York: HarperCollins, 2004.

Jehl, Francis. *Menlo Park Reminiscences.* Mineola, N.Y.: Dover, 1990.

Kaku, Michio. *Einstein's Cosmos: How Albert Einstein's Vision Transformed Our Understanding of Space and Time.* New York: Norton, 2004.

Kevles, Daniel J. *The Baltimore Case.* New York: Norton, 1998.

Kragh, Helga. "Max Planck: The Reluctant Revolutionary," *Physics World,* December 2000.

Krauss, Lawrence. *The Physics of Star Trek.* New York: HarperCollins/Basic Books, 1995.

Kuhn, T. S. *The Copernican Revolution.* Cambridge, Mass.: Harvard University Press, 1957.

Lerner, Eric J. "What's Wrong with the Electric Grid?" *Industrial Physicist*, October–November 2003.

Lewis, C. S. *The Discarded Image: An Introduction to Medieval and Renaissance Literature*. Cambridge: Cambridge University Press, 1964.

Lindley, David. *Where Does the Weirdness Go? Why Quantum Mechanics Is Strange, But Not as Strange as You Think*. New York: HarperCollins/Basic Books, 1996.

Livio, Mario. *The Accelerating Universe: Infinite Expansion, the Cosmological Constant, and the Beauty of the Cosmos*. New York: Wiley, 2000.

———. *The Golden Ratio*. New York: Broadway, 2002.

Ludwig, Charles. *Michael Faraday, Father of Electronics*. Scottdale, Pa.: Herald Press, 1978.

Mahon, Basil. *The Man Who Changed Everything: The Life of James Clark Maxwell*. New York: Wiley, 2003.

Mann, Alfred K. *Shadow of a Star: The Neutrino Story of Supernova 1987A*. New York: W. H. Freeman, 1997.

Mayfield, Kendra. "E-Ink: Your Hands Will Thank You," *Wired.com*, March 1, 2001.

Mean Girls. Dir. Mark S. Waters, 2004, Paramount.

Michelmore, Peter. *The Swift Years: The Robert Oppenheimer Story*. New York: Dodd Mead, 1969.

Miesel, Sandr. "Dismantling the Da Vinci Code," *Crisis Magazine*, September 2003.

Moore, Alan. *The League of Extraordinary Gentlemen*, Vol. I. La Jolla, Calif.: America's Best Comics, 1999.

Moran, Richard. *Executioner's Current: Thomas Edison, George Westinghouse, and the Invention of the Electric Chair*. New York: Knopf, 2002.

My Big Fat Greek Wedding. Dir. Joel Zwick, 2002, Gold Circle Films.

Neeley, Kathryn A. *Mary Somerville: Science, Illumination, and the Female Mind*. Cambridge: Cambridge University Press, 2001.

Neutrinos and Beyond: Windows on Nature. National Research Council report, 2003.

Ocean's Eleven. Dir. Steven Soderbergh, 2001, Warner Brothers.

Olson, Lynn M. *Women in Mathematics*. Cambridge, Mass.: MIT Press, 1974.

Ouellette, Jennifer. "Pollock's Fractals," *Discover*, November 2001.

———. "Teaching Old Masters New Tricks," *Discover*, December 2001.

———. "Bubble, Bubble: The Physics of Foam," *Discover*, June 2002.

Pandolfini, Bruce. *Kasparov and Deep Blue: The Historic Chess Match between Man and Machine*. New York: Fireside, 1997.

Park, Robert L. *Voodoo Science*. Oxford: Oxford University Press, 2000.

Parker, Steve. *How Things Work*. New York: Random House, 1991.

Parnell, Peter. *QED*. New York: Applause Theatre and Cinema Books, 2002.

Pasachoff, Naomi. *Alexander Graham Bell: Making Connections*. Oxford: Oxford University Press, 1996.

Pears, Iain. *An Instance of the Fingerpost*. New York: Riverhead, 1998.

Penrose, Roger. *The Emperor's New Mind: Concerning Computers, Minds, and the Laws of Physics*. Oxford: Oxford University Press, 1990.

Perkowitz, Sidney. *Empire of Light*. Washington, D.C.: Joseph Henry, Press, 1996.

———. *Universal Foam*. New York: Walker, 2000.

Permutter, Saul. "Supernovae, Dark Energy, and the Accelerating Universe," *Physics Today*, April 2003.

Peterson, Ivars. *Newton's Clock: Chaos in the Solar System*. New York: W. H. Freeman, 1993.

Pleasantville. Dir. Gary Ross, 1998, New Line Cinema.

Poe, Edgar Allan. "The Purloined Letter." In *Selected Prose, Poetry, and Eureka*. Austin, Tex.: Holt Rinehart and Winston.

Pullman, Philip. His *Dark Materials*. New York: Random House/Knopf Books for Young Readers, 2000. (Boxed set.)

Quantum Universe: The Revolution in Twenty-First-Century Particle Physics. Report of the Department of Energy/National Science Foundation High Energy Physics Advisory Panel, 2004.

Rashomon. Dir. Akira Kurosawa, 1950, Daiei Studios (Japan).

Regis, Ed. *Nano: The Emerging Science of Nanotechnology*. New York: Little Brown, 1995.

Rhodes, Barbara J., and William W. Streeter. *Before Photocopying: The Art and History of Mechanical Copying, 1780 to 1938*. New Castle, Del: Oak Knoll Press, 1999.

Rhodes, Richard. *The Making of the Atomic Bomb*. New York: Simon and Schuster, 1986.

———. *Dark Sun: The Making of the Hydrogen Bomb*. New York: Simon and Schuster, 1996.

Rincon, Paul. "Nanotech Guru Turns Back on 'Goo,' " *BBC News Online*, June 9, 2004.

Samuel, Eugenie. "Rising Star of Electronics Found to Have Fabricated His Data," *New Scientist*, October 5, 2002.

Sayers, Dorothy. *Gaudy Night*. New York: HarperCollins/Avon, 1968.

Schiffman, Betsy. "Current Costs," *Forbes.com*, March 2004.

Schwarz, Cindy, and Sheldon Glashow. *A Tour of the Subatomic Zoo: A Guide to Particle Physics*. Woodbury, N.Y.: American Institute of Physics Press, 1996.

Schweber, Silvan S. *QED and the Man Who Made It*. Princeton, N.J.: Princeton University Press, 1994.

Searle, John R. *Minds, Brains, and Action.* Cambridge, Mass.: Harvard University Press, 1984.

Segre, Gino. *A Matter of Degrees: What Temperature Reveals about the Past and Future of Our Species, Planet, and Universe.* New York: Penguin, 2002.

Seife, Charles. *Zero: The Biography of a Dangerous Idea.* New York: Penguin, 2000.

Shakespeare, William. *Troilus and Cressida.* Riverside Shakespeare. Boston, Mass.: Houghton Mifflin, 1974.

Shermer, Michael. *Why People Believe Weird Things: Pseudoscience, Superstition, and Other Confusions of Our Time.* New York: W. H Freeman, 1997.

Singh, Simon. *The Code Book.* New York: Random House/Anchor, 1999.

Sleeper. Dir. Woody Allen, 1973, MGM/United Artists.

Snow, C. P. *The Search.* New York: Scribner, 1934.

Steadman, Philip. *Vermeer's Camera.* Oxford: Oxford University Press, 2001.

Stepford Wives, The. Dir. Bryan Forbes, 1975, Anchor Bay Entertainment.

Stern, Philip M. *The Oppenheimer Case: Security on Trial.* New York: Harper-Collins, 1969.

Stevenson, Robert Louis. *Dr. Jekyll and Mr. Hyde.* New York: Bantam, 1981

Swade, Doron. *The Difference Engine: Charles Babbage and the Quest to Build the First Computer.* New York: Penguin, 2002.

Szasz, Ferenc M. *The Day the Sun Rose Twice: The Story of the Trinity Site Nuclear Explosion, July 16, 1945.* Albuquerque: University of New Mexico Press, 1995.

Terminator 2: Judgment Day. Dir. James Cameron, 1991, Artisan Entertainment.

Tillyard, E. M. W. *The Elizabethan World Picture.* New York: Random House/Vintage Books, 1959.

Time After Time. Dir. Nicholas Meyer, 1979, Warner Brothers.

Tolkien, J. R. R. *The Lord of the Rings.* New York: Ballantine, 1965. (Trilogy.)

Turner, Michael. "Plucking the Strings of Relativity," *Discover*, September 2004.

Updike, John. "Cosmic Gall." In *Telephone Poles and Other Poems.* New York: Random House/Knopf, 1959.

Verne, Jules. *From the Earth to the Moon.* New York: Bantam, 1993.

Visitors, The. Dir. Jean-Marie Poire, 1996, Miramax.

Wachhorst, Wyn. *Thomas Alva Edison: An American Myth.* Cambridge, Mass.: MIT Press, 1981.

Wasserman, Harvey. "California's Deregulation Disaster," *Nation*, February 12, 2001.

Waters, Roger. "Radio Days." In *Radio K.A.O.S.* Sony, 1987.

Weiss, Richard J., ed. *The Discovery of Antimatter: The Autobiography of Carl David Anderson, the Youngest Man to Win the Nobel Prize.* Hackensack, NJ.: World Scientific, 1999.

Wells, H. G. *The Time Machine.* New York: Airmont, 1964.

———. *War of the Worlds.* New York: Pocket Books, 1988.

Westfall, Richard S. *Never at Rest: The Life of Isaac Newton.* Cambridge: Cambridge University Press, 1980.

Whitbeck, Caroline. "Trust and the Future of Research," *Physics Today*, November 2004.

Whitfield, John. "Accelerating Universe Theory Dispels Dark Energy; Tweaking Gravity Does Away with Need for Strange Forces," *Nature*, July 3, 2003.

Williams, Jay. *Leonardo da Vinci.* New York: American Heritage, 1965.

Williams, L. P. *Michael Faraday: A Biography.* New York: Basic, 1965.

Willis, Connie. *The Doomsday Book.* New York: Bantam, 1992.

———. "At the Rialto." In *Impossible Things.* New York: Bantam Spectra, 1994.

———. *To Say Nothing of the Dog.* New York: Bantam, 1998.

Winterson, Jeanette. *Gut Symmetries.* New York: Random House/Vintage International, 1998.

Wite, Lynn, Jr. "Eilmer of Malmesbury: An Eleventh-Century Aviator," *Technology and Culture* 2, no. 2, Spring 1961.

Woolley, Benjamin. *The Bride of Science.* New York: McGraw-Hill, 1999.

Woosnam, Maxwell. *Eilmer: Eleventh-Century Monk of Malmesbury.* Malmesbury: Friends of Malmesbury Abbey, 1986.

Worrall, Simon. *The Poet and the Murderer.* New York: Penguin, 2002.

X: The Man with X-Ray Eyes. Dir. Roger Corman, 1963, MGM/United Artists.

X-Files, The. "D.P.O." Season 3. Dir. Kirn Manners. Written by Howard Gordon. Air date: October 6, 1995. Twentieth Century Fox.

———. "Kill Switch." Season 5. Dir. Rob Bowman. Written by William Gibson and Tom Maddox. Air date: February 15,1998.

X-Men. Dir. Bryan Singer, 2000, Twentieth Century Fox.

ONLINE RESOURCES

Adventures in Cybersound: *http://www.acmi.net.au/AIC*

Airship Heritage Trust: *www.aht.ndirect.co.uk*

Albert Einstein Institute: *www.aeinstein.org*

Alexander Graham Bell Institute: *http://bell.uccb.ns.ca*

American Institute of Physics Online Exhibits: *www.aip.org/history*

Amusement Park Physics: *www.learner.org/exhibits/parkphysics/coaster.html*

Biomimicry Project: *www.biomimicry.net*

Centennial of Flight: *www.centennialofflight.af.mil*

Century of Radiology: *www.xray.hmc.psu.edu/rci/ss1/ss1_2.html*

Charles Babbage Institute: *www.cbi.umn.edu*

Coaster Central: *www.coastercentral.com*

Computing History Museum: *www.computinghistorymuseum.org*

Copernicus Museum in Frombork: *www.frombork.art.pl*

Edison Birthplace Museum: *www.tomedison.org*

Edison Home Page: *www.thomasedison.com*

Feynman Online: *www.feynmanonline.com*

Franklin Institute: *http://sln.fi.edu*

Galileo Project: *http://galileo.rice.edu*

Grado, Victor M. "Nepohualtzitzin, A Mesoamerican Abacus," March 28, 1999: *www.ironhorse.com/~nagual/abacus*

History of Camera Lucida: *www.cameralucida.org.uk/*

History Channel: *www.historychannel.com*

How Stuff Works: *www.howstuffworks.com*

Hubble Telescope: *http://hubble.stsci.edu*

Leggat, Robert. "History of Photography": *www.rleggat.com/photohistory*

Leonardo da Vinci Museum: *www.leonardo.net*

Live from CERN: *http://livefromcern.web.cern.ch/livefromcern*

Lucent Technologies/Bell Laboratories Investigative Committee Report: *www. lucent. com/news_events/pdf/researchreview.pdf*

Marconi Heritage Foundation: *www.gec.com/about_marconi/heritage*

Maxwell's Demon: *www.maxwelliandemon.co.uk*

Molecular Expressions: *http://micro.magnet.fsu.edu*

Morse Society: *www.morsesociety.org*

Museum of Science: *www.mos.org*

National Inventor's Hall of Fame: *www.invent.org*

National Museum of Science and Technology: *www.museoscienza.org/english*

Newton Institute for Mathematical Sciences: *www.newton.cam.ac.uk*

Newton Project: *www.newtonproject.ic.ac.uk*

Nikola Tesla Museum: *www.yurope.com/org/tesla/index.htm*

Nobel Prize Foundation: *www.nobel.se/physics/laureates*

Oppenheimer Hearing Transcript: *www.yale.edu/lawweb/avalon/abomb/opp06.htm*

Pepper's Ghost: *www.dafe.org/misc/peppers/peppers.htm*

Physics Central: *www.physicscentral.com*

Roentgen Centennial: *http://web/wn/net/ricter/web/history.html*

Robert Hooke: *www.roberthooke.org.uk*

San Francisco Exploratorium: *www.exploratorium.edu*

Skeptical Inquirer, The: *www.csicop.org/si*

St. Andrews History of Mathematicians: *http://www-gap.dcs.st-and.ac.uk/*

~history/mathematicians

Tesla Memorial Society of New York: *www.teslasociety.org*

Today in Science: *www.todayinsci.com*

Trinity Test: *http://nuketesting.environweb.org* (videos, photos, documents); *www.dannen.com/decision/trin-eye.htm* (eyewitness accounts); *www.nuclearfiles.org/redocuments/docs-trinity.html* (documents)

Ultimate Roller Coaster: *www.UltimateRollerCoaster.com*

University of Dundee X-Ray Museum: *www.dundee.ac.uk/museum*

Velcro: *www.velcro.com*

Very Brief History of Copying: *www.sit.wisc.edu/~sandness/history.html*

Virtual Museum of History of the Telescope: *www.polarisinteractive.com/vmht*

Virtual Museum of Isaac Newton: *www.newton.org.uk*

Wikipedia Encyclopedia: *www.wikipedia.com*

William Herschel Museum: *www.bath-preservation-trust.org.uk/museums/herschel*

William Herschel Society: *www.williamherschel.org.uk*

X-Ray Century: *www.emory.edu/X-RAYS/century.htm*

Zeppelin Museum: *www.zeppelin-museum.de/firstpage.en.htm*

INTERACTIVE

Build Your Own Roller Coaster: *www.fearofphysics.com/roller/roller.html*; *http://dsc/discovery.com/convergence/coasters/interactive/interactive.html*

Chaos Game: *http://math.bu.edu/DYSYS/chaos-game/node1.html*

Electromagnetic induction: *http://micromagnet.fsu.edu/electromag/jaza/faraday2*

Electromagnetic rotation/dynamo: *www.phy6.org/earthmag/dynamos.htm*

Film of horse galloping: *http://freeweb.pdq.net/headstrong.zokit.htm*

Maxwell's Demon: *http://ajs.net/maxwell.htm*; *www.chem.uci.edu/education/undergrad_pgm/applets/bounce/demo.htm*

Paper airplane aerodynamics: *http://paperplane.org*

Particle Adventure: *http://particleadventure.org*

Potential/Kinetic Energy: *http://zebu.uoregon.edu/1998/ph101/pel.html*

Relativistic Flight through Stonehenge: *www.itp.uni-hannover.de/~dragon/stonehenge/stoneng.htm*

Schrödinger's Cat: *www.phobe.com/s_cat/s_cat.html*

Top Quark Game: *http://education.jlab.org/topquarkgame*

Acknowledgements

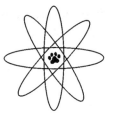

This book would not have been possible without the encouragement and support of Alan Chodos of the American Physical Society. Much of the impetus for bringing this project to fruition is the result of his gentle yet insistent prodding, and I am extremely grateful to him for it.

I am equally indebted to my agent, Mildred Marmur, and to Jane von Mehren and Caroline White at Viking/Penguin, who recognized the book's potential. Thanks to Caroline's keen editorial eye, the prose is mercifully free of mixed metaphors, shaky analogies, lapses in logic, and just plain awkward sentence structure.

Ann Kottner proved a diligent and thorough research assistant, digging up tons of quirky historical nuggets. Peri Lyons Thalenberg, Bob Mondello, and Carlos Schröder endured countless conversations about random physics concepts, and suggested numerous historical, literary and pop-culture trivia to help illustrate them. Like me, they suffer from chronic "magpie mind"—both a blessing and a curse. The D.C.-based Roomful of Writers group bravely plowed through several of the draft chapters, despite the scary physics content, and provided thoughtful critiques that greatly improved the finished product.

I am fortunate to count numerous physicists and science writers among my friends and acquaintances, many of whom took the time to explain concepts, peruse chapters, or simply talk through the logistical difficulties of translating abstractions into everyday English: Kenneth Chang, the aforementioned Alan Chodos, Jessica Clark, Charles Falco, Susan Ginsberg, Martha Heil, Bob Park, Sidney ("El Sid") Perkowitz, James Riordon, Brian Schwartz, Francis Slakey, and Ben Stein. Their combined expertise was especially helpful in guiding me through the murky waters of physics theory, where I often found myself in a bewildering superposition of states: being not quite right, and not quite wrong. If the book has any remaining failings in those areas, they are mine alone.

No writer can function in a vacuum, devoid of friends and family. My heartfelt appreciation goes out to those who suffer my many foibles without (much) complaint, including Frank and Cecelia Brescia; Adam Cvijanovic; Amy English; Gina Eppolito; Larry Feinberg; Erica Friedman; Kat James; Rich Kim; Brian Mc-Cormick; Dave Olson; David and Rachel Ouellette; Alison Pride; Shari Steelsmith Duffin; Don Wagner; and the collective folks at Bay Ridge Dojo—a large part of my heart will always be in Brooklyn because of all of you. Very special thanks go to my former college English professor, Janet Blumberg, whose passion for knowledge, intellectual rigor, and gift for finding quirky connections indelibly shaped my thinking.

Last, I will always be grateful to my parents, Paul and Jeanne, for encouraging me to go out into the cold, cruel world and carve out my own idiosyncratic path—one vastly different from theirs in almost every conceivable way, yet they love me not one whit less. And vice versa.

Index

abaci 67
AC electricity 111–14
acceleration, law of 38
action, equal and opposite reaction for 38, 74, 166, 247
Action Group on Erosion, Technology, and Concentration (ETC) 230
Addams Family Values 34, 37
aerodynamic theory 74–7
aerosol cans, invention of 203
air, as fluid 74
Alcoke, Charles (roller coasters) 105
Aldrin, Lunar Module Pilot Edwin "Buzz" 171
Alhazen (11th-century Arab scholar) 7, 51–2
Amber Spyglass, The (Philip Pullman) 27
American Gods (Neil Gaiman) 9
"analytical engine" 70–1
Anderson, Carl (antiparticle) 240
animal magnetism (mesmerism) 60–1
antimatter 238–40, 242–4, 246–7
Apfel, Robert (handheld portable neutron detector) 206
Apollo 11 132, 171–2
Apollo 13 66
Apollo's Chariot (roller coaster) 102
Archer, Frederick Scott (photography) 122
Aristotle (philosopher) 149
Armitage, Karole (*The Schrödinger Cat* ballet) 174
Armstrong, Commander Neil A. 171
art: fractal geometry in 235; optical effects and 50; theory of optical aids used in 50, 56; use of camera lucida in 54–6; use of camera obscura in 50–6
artificial emotion 258
artificial intelligence (AI) 253–9
"At the Rialto" (Connie Willis) 173–5, 179–80
Atalay, Bulent (*Math and the Mona Lisa*) 3
atomic bombs 188–194
atoms 246–7; *see also* subatomic particles
Auburn, David (*Proof*) 87
Austin Powers: International Man of Mystery 218, 221

Babbage, Charles (calculating machine) 67–72, 92
Back to the Future 157, 163
bacteria, early observations of 23, 25
ballooning 77
Batlogg, Bertram (Schön fraud) 276
Batman Forever 147
battery, invention of 48–9
Belin, Edouard (first telephoto transmissions) 144
Bell, Alexander Graham: flight experiments of 79; photophone invention 144; telephone invention 97–8, 101; use of biomimicry 212
Benyus, Janine (*Biomimicry: Invention Inspired by Nature*) 211
Berginger, Johann (Gensteine fraud) 273
Bernoulli, Daniel (aerodynamic theory) 74–6
Bernoulli principle 74–6
big bang, antimatter and 246

biomimicry 210–12

Biomimicry: Invention Inspired by Nature (Janine Benyus) 211

black holes 282

blackbody, thermodynamics and 152

blackouts 114–17

Blade Runner (Ridley Scott) 252, 259

Blanchard, John (balloon) 77

Bleriot, Louis (flight) 79

Blondlot, René (N ray fraud) 273

Bohr, Niels (particle-wave duality) 177

Bolor, Denis (flight) 73

Book of Calculation (Fibonacci) 3

Boreel, William (Dutch diplomat) 21

Bose, Satyendra (Bose-Einstein condensate) 222

Bose-Einstein condensate (BEC) 222–3

bosons 247

Bradbury, Ray ("I Sing the Body Electric") 257–8

Brahe, Tycho (Copernicus critic) 13–14

brain, function of 253

Bringsjord, Selmer (Brutus. 1 creator) 257–8

Brooks, Rodney (Kismet robot creator) 258–9

Brown, Dan (*The da Vinci Code*) 1–3

Bruno, Giordano (scientific martyr) 15

bubbles, *see* foam

Buffy the Vampire Slayer 234

"butterfly effect" 233–4

Butterfly Effect, The 233

calculating machines, invention of 67–8

calotype 122

camera lucida 54–6

camera obscura 50–6, 118–21; *see also* spectroscopic projection

Cameron, James (*Terminator 2: Judgment Day*) 192

Campbell, Murray (Deep Blue creator) 255

Canton, John (lightning) 45

Čapek, Karel (*R.U.R.*) 259

Carlson, Chester (copy machine) 183–6

Carnot, Sadi (thermodynamics) 83–4

"Carnot cycle" 84

cathode-ray tube (CRT) 142

Catholic church, *see* Christian theology

Cayley, Sir George (aerodynamic theory) 74

chaos theory 231–2, 234; *see also* fractal geometry

Charlie and the Chocolate Factory (Roald Dahl) 142

Cheney, Margaret 112

chess, automating 254–9

"Chinese room" parable 254

Christian theology: creationism and 16–17; criticism of *De revolutionibus* and 14–15; embracing Copernican system and 17; Galileo and 15–16, 29; martyrdom of Giordano Bruno and 15; Ptolemy's cosmological model and 10

chromatic aberration 21

Chu, Steven (laser cooling) 222

CO_2 lasers 221–2

Codex Atlanticus (Leonardo da Vinci) 7

Collins, Command Module Pilot Michael 171

color, light experiments and 149–50

compass, invention of simple 60–1

Compton, Arthur (light) 155

conductance 43

Coney Island roller coasters 105

Cooke, William F. (telegraph development) 96

Copernicus, Nicolaus: creating/publishing *De revolutionibus* 12–13; criticism of theory of 13–15; effect of *De revolutionibus* 10; launching scientific revolutions 8

copy machine, invention of 185–7

copying, history of 184; *see also* photocopying

Cornell, Eric (BEC) 222

cosmic microwave background radiation 242–3

cosmic rays, antiparticles in 247–8

cosmological models 10–12

counting devices 67

Cowan, Clyde (neutrino) 268

creationism support of 16–17

Crichton, Michael: *Jurassic Park* 231; *Prey* 224, 228–9, 258–9

Crippen, Dr. Hawley (murderer) 129–30

Crooke, Sir William (Crooke's tube) 142

Crooke's tube 136, 142

Ctesibius (Greek) 212

cure, magnetism as 60–1, 65

Cvijanovic, Adam (artist) 51

da Vinci, Leonardo: camera obscura and 52; flight and 73; foam studies 207; life and art of 5–8; as scientist/inventor 7–8; use of biomimicry 212

da Vinci Code, The (Dan Brown) 1–3

Daguerre, Louis (photographer) 121

daguerreotypes 122

Dahl, Roald (*Charlie and the Chocolate Factory*) 142

Dalí, Salvador (artist) 3

d'Alibard, Thomas François (lightning) 45

Dally, Clarence (Edison's assistant) 138

dark energy 263–5

dark matter 260–1, 264–5

Darwin, Charles (evolution theorist) 16

Davis, Ray, Jr. (neutrino) 268–9

DC electricity 111

de Broglie, Louis (particle-wave duality) 176

De divina proportione (Leonardo da Vinci) 5

de la Roche, Charles François Tiphaigne (*Giphantie*) 120

de Mestral, George (Velcro) 213–14

De revolutionibus 12–15

de Saint-Victor, Abel Niepce (photography) 122

de Villanova, Arnau (13th-century magician) 119

Deep Blue (computer) 256–7

Deep Thought (computer) 255

della Porta, Giambattista (16th-century Italian scientist) 52, 118

Delor, M. (lightning) 45

deregulation, of electrical industry 115–17

Deroche, Elise (licensed female pilot) 79

"Dialogue concerning the Two Chief World Systems" (Galileo Galilei) 16

Dicke, Robert (cosmic microwave background radiation) 244

"difference engine" 69–71

Difference Engine, The (William Gibson, Bruce Sterling) 71

Dircks, Henry (spectroscopic projection) 119

dirigibles, development of 78–9

Disneyland roller coasters 105–6

divine proportion 2–4

Doomsday Book, The (Connie Willis) 164

Doppler, Christian ("Doppler shift") 262

"Doppler shift" 262

Dr. Jekyll and Mr. Hyde (Robert Louis Stevenson) 148, 156

Dr. Strangelove, or How I Learned to Stop Worrying and Love the Bomb (Stanley Kubrick) 188, 191, 194

drag 76, 166–8

Drexler, Eric (*Engines of Creation*) 224, 230

du Chatelet, Émilie 89–90

dual personality, in popular fiction 148–9

Durian, Douglas (foam) 209

Eastgate Building, biomimicry of 211–12

Edison, Thomas: inventing motion pictures 122–5; mimeograph invention 184; Tesla and 112–14

Eilmer of Malmesbury (flight) 73

Einstein, Albert: $E=mc^2$ equation and atomic bombs 188–9; expansion of universe and 262–4; extending Planck's quanta 155; gravity studies of 40–1; life of 158–9; stimulated emission studies 218–20; theories of special relativity 158–65 electricity: blackouts 114–17; deregulation of industry 115–17; development of AC 111–14; invention of battery and 48–9; lightning and 44–7; magnetism and 59–65; overview 43–5; types of current 111

electrocution, AC power and 113–14

electromagnetic rotation 62

electromagnetism 59–65, 280–1; *see also* magnetism

electronic paper, photocopying and 187

electrons: creating electricity and 44; light creation and 19; particle-wave duality studies 176–80; QED theories and 200–2

electrophotography 185

Elegant Universe, The (Brian Greene) 284–5

element 118 fraud 274–6

energy: dark 263–5; potential versus kinetic 37–8; understanding 83–4

Engines of Creation (Eric Drexler) 224

entropy 82

Escher, M. C. (artist) 83

ETC, *see* Action Group on Erosion, Technology, and Concentration (ETC)

Euler, Leonard (18th-century Swiss mathematician) 282

evolutionary robotics 258

Fabio (model) 102,107

Falco, Charles (physicist) 56–8

Faraday, Michael (electromagnetism) 62–4

Farnsworth, Philo (television) 144–5

Fellowship of the Ring, The (J. R. R. Tolkien) 225

Fermi, Enrico (neutrino) 267

Fermi National Laboratory 248

fermions 246

Ferucci, Dave (Brutus. 1 creator) 257

Feynman, Richard: atomic bomb results and 192; on nanotechnology future 229–30; QED studies 196–202

Feynman diagrams 200–1

Fibonacci sequence 3

Fisher, Frederick (ghost story) 120

flight 74–80

Fludd, Robert (perpetual motion waterwheel) 81–3

"flugtag" competitions 76

foam 203–9

forces, four fundamental 283–4

forgery 182, 187

fractal geometry 234–5; *see also* chaos theory

Franklin, Benjamin: early years of 44–5; invention of lightning bells 47; invention of lightning rod 47–8; lightning experiments of 45–6

fraud: difficulties uncovering 277–8; forgery 182, 187; overview 272–5; of Schön, 275–7 *From the Earth to the Moon* (Jules Verne) 166, 169

Frost, Edwin Brant (X rays) 137

fuel: antimatter as 242; rocket 168–9

g forces, roller coasters and 107

Gaiman, Neil (*Good Omens*) 94

Galilei, Galileo: experiments with gravity 34–6; observations with telescope 28–9; supporting Copernicus 15–16

Galvani, Luigi (electricity) 48–9

Gates, James (string theorist) 285

Gaudy Night (Dorothy Sayers) 272–3, 278

Gell-Mann, Murray (physicist), on crisis of physics 196

Gemma-Frisium, Reinerus (Dutch scientist) 52

geometry, fractal 234–5

Germain, Sophie (French mathematician) 90–2

Gibson, William (*Difference Engine, The*) 71

Gibson, William (*The X-Files*) 252

Giffard, Henri (dirigible) 77

Gilbert, William (electricity) 42

Gingerich, Owen (*The Book Nobody Read*) 10

Giphantie (Charles François Tiphaigne de la Roche) 120

Goddard, Robert H. (liquid-fuel rocket) 169–72

Golden Compass, The (Philip Pullman) 81, 260

golden ratio 2–4

Good Omens (Neil Gaiman, Terry Pratchett) 94

Gopala (ancient Indian scholar) 3

Gould, Gordon (laser) 220

grand unified theory (GUT) 279–80, 283

gravity: Einstein's studies of 40–1; Galileo's experiments with 34–6; Newton's studies of 37–40; as physical force 281; roller coasters and 103, 107

Gravity Pleasure Switchback Railway 105

Greenbaum, Elias (photosystem I) 214–15

Greene, Brian (*The Elegant Universe*) 284–5

Gribbin, John (particle-wave duality) 176

GUT (grand unified theory) 279–80, 283

Gut Symmetries (Jeanette Winterson) 174–5, 189, 279, 293

Gutenberg, Johannes (printing press) 183

Harry Potter books 187

Heaviside, Oliver (electromagnetism) 64

hectograph copy machine 184

Heisenberg, Werner (particle-wave duality) 177

heliocentric Copernican cosmological model 11, 14–15

Herbert, Mary Sidney, countess of Pembroke (Elizabethan chemist) 89

Herschel, William (astronomer) 30–32

Hertz, Heinrich (electromagnetism) 64

Higgs boson 250–1

Highsmith, Roger (*The Science of Harry Potter*) 187

Hindenburg (dirigible) 78

Hinckle, Phillip (roller coasters) 105

Hiroshima, bombing of 191–2

Hockney, David (artist) 50, 56–8

Hoffman, Mark (forger) 181–2, 186–7

Holt, Glynn (foam) 209

Hooke, Robert (*Micrographia*) 22–3

Hopkins, Gerard Manley (Victorian poet) 216

Hsu, Feng Hsiung (Deep Blue creator) 255–6

Hubble, Edwin, expansion of universe and 262–3

Hubble space telescope 26–7, 32–3

hydrogen bomb 193

Hypatia (4th-century female scientist) 88

"I Sing the Body Electric" (Ray Bradbury) 257–8

Ice Storm, The 43

Imanishi-Kari, Teresa (fraud accusations) 274

Independence Day 218

inertia, law of 37–8

injuries, on roller coasters 107–8

Inquisition, Galileo and 16, 29

interplanetary communication 132–3

ions, electricity and 43

iridescent, studies of 219

It's a Wonderful Life 234

Jackson, Peter (*Lord of the Rings* movie trilogy) 126

James Webb space telescope 33

Janssen, Zacharias (microscope) 21–2

jet propulsion 166–7

Joliot-Curie, Frédéric (atomic energy) 189

Joliot-Curie, Irène (atomic energy) 189

Jung, Carl (psychologist) 80

Jurassic Park (Michael Crichton) 231

Kasparov, Gary (chess champion) 255–7

Kemmler, William, electrocution of 113

Kepler, Johannes: camera obscura use 51; laws of planetary motion 14

Ketterle, Wolfgang (BEC) 223
kinetic energy, versus potential 37–8
Kismet robot 258
Kornei, Otto 185
Kraynik, Andrew (foam) 207–8
Kubrick, Stanley (*Dr. Strangelove or How I Learned to Stop Worrying and Love the Bomb*) 188, 191, 194

lambda 261–2, 264
Lapin, Aaron "Bunny" (Reddi-Wip creator) 203, 205
Large Hadron Collider (LHC) 250
lasers 217–23
law of inertia 37–8
laws of motion 37–9, 166–7
laws of thermodynamics 82–5
Le Corbusier (French architect) 3
League of Extraordinary Gentlemen, The (Alan Moore) 149
length contraction 160–4
lens 19–22
leptons 246, 248, 251
Leviathan telescope 31
Leyden jar 45–7
light: dual nature of 149, 155–6; early experiments with 149–51; extending Planck's quanta to 153–6; in lasers 217–23; magnetism and 63; motion and 160–1; properties of 18–19; refraction of 19–20
lightning 42, 44–7
lightning bells invention 47–8
lightning rod invention 47–8
Lippershey, Hans (telescope) 27–8
literature, produced by artificial intelligence 253–9; *see also* specific titles "Little Commentary" (*Commentariolus*) 12, 14–15
Livio, Mario (*The Golden Ratio*) 3
Lomonosov, Mikhail (lightning) 45
Lord Kelvin's cell 208
Lord of the Rings movie trilogy 126
Lorenz, Edward (chaos theory) 231–2
Lotto, Lorenzo (artist), potential use of optical aids 56

Lovelace, Ada (British mathematician/scientist) 92
Luther, Martin (Copernicus critic) 14–15
Lutter, Vera (artist) 51, 121

McEuen, Paul (Schön fraud discovery) 275
Machack IV computer 255
Maddox, Richard (photography) 122
Magnes (Greek shepherd) 59
magnetic resonance imaging (MRI) 65
magnetism 49, 59–65
magnifying stones 18
Manhattan Project 190, 193
Man Out of Time (Margaret Cheney) 112
Marconi, Guglielmo (radio invention) 128–9, 131–3
mass, gravity and 35
Mathematical Principles of Natural Philosophy (*Principia*) (Isaac Newton) 39–40
mathematics: fractal geometry 234–5; women in 87–8, 90–3
matra-vrttas 3
Matrix, The 253
Matzke, Edwin (foam) 207–8
Maxwell, James Clerk: electromagnetism studies 63–4; light experiments 151; thermodynamics studies 85–6
"Maxwell's demon" 85–6
Mean Girls 87, 93
medicine, X rays in 137–9
Mesmer, Franz Anton (mesmerism) 60–1
Meucci, Antonio (telephonic device) 100–1
Micrographia (Robert Hooke) 23–4
microscope 21–5
mimeograph copy machine 184
molecular spectroscopy 219–20
Mona Lisa (Leonardo da Vinci) 1, 3–6
Mongolfier brothers (balloon) 77
Moore, Alan (*The League of Extraordinary Gentlemen*) 149

Morse, Samuel Finley Breese (telegraph) 97

motion pictures 122–5

movies; *see* specific movies

music, fractal geometry in 235–6

Muybridge, Eadweard (photography) 122–3

My Big Fat Greek Wedding 245–6

N ray fraud 273

Nagasaki, bombing of 189, 191–2

nanobots 224, 227–8

nanotechnology 224–30

neutrinos 266–71

Newton, Isaac: early life of 36; gravity studies of 36–7; light experiments 149–51; physical laws of 37–9; *Principia* 39–40; reflecting telescope and 29–30

Niepce, Joseph-Nicóphore (photographer) 121

Ninov, Victor (element 118 fraud) 275–6

nuclear fission, atomic bomb and 189–90

nuclear force 280–1

"Oath of a Freeman" forgery 181–2

Oberth, Hermann (*The Rocket into Interplanetary Space*) 169

Ocean's Eleven 64–5

Oppenheimer, J. Robert (father of atomic bomb) 190, 193–4

Ørsted, Hans Christian (magnetism) 61–2

Osiander, Andreas (*De revolutionibus* publisher) 13

Our Town (Thornton Wilder) 268

paradox of Schrödinger's cat 174

Parnell, Peter (QED) 199, 202

Parsons, William (Lord Rosse) (Leviathan telescope) 31

Particle Data Group 249

particle physicists, role of 248–9

particle-wave duality 176–80

particles: quantum mechanics and 176–80; subatomic 245–51, 265–8

Pauli, Wolfgang (neutrino) 267

Pearce, Mick (architect) 211–12

Peary, Robert (Arctic explorer) 129

Pennell, Joseph (American etcher and lithographer), on Vermeer 52–3

Penzias, Arno (cosmic microwave background radiation) 243–4

People Like Us 74

Pepper, "Professor" Henry (spectroscopic projection) 119–20

"Pepper's Ghost" (spectroscopic projection) 119–20

Pepys, Samuel (diarist) 23

perception, paradox of Schrödinger's cat and 174

Peregrinus, Petrus (compass) 59–60

Perkowitz, Sidney (Universal Foam) 205–6

perpetual motion machines 82–4

Phelan, Robert (foam) 208

phi 2–6

photo conductivity 184–5

photocopying 185

photography 118–26; *see also* camera lucida; camera obscura

photonic crystal, studies of 216

photons: in lasers 218–20; light and 19; particle-wave duality studies 176–80; QED theories and 200–2; X-ray 134–5

photosystem I (PSI) studies 214–15

physics, division of 280

Pi Sheng (printing press) 182

pinhole camera 120

Pisa, Leonardo of (Fibonacci) 3

Planck, Max (thermodynamics) 151–5

Plateau, Joseph (foam) 207

Pleasantville 141, 147

Poe, Edgar Allan ("The Purloined Letter") 266–7, 269–71

Pollock, Jackson (artist) 235, 237

positron 241

potential energy, versus kinetic 37–8

power outage, *see* blackout

Pratchett, Terry (*Good Omens*) 94–5

Prey (Michael Crichton) 224, 228–9, 258–9

Prince of Wales, Charles 224

printing, history of 181–3
prisms 54–5, 149–50
Proof (David Auburn) 87
PSI (photosystem I) studies 214–15
Ptolemy (Almagest), cosmological model of 10–12
Pullman, Philip: *The Amber Spyglass* 27; *The Golden Compass* 81, 260; *The Subtle Knife* 260–1
"Purloined Letter, The" (Edgar Allan Poe) 266–7, 270
Pythagoras 149

QED (Peter Parnell) 199, 202
quantum communication 223
quantum electrodynamics (QED) 196–202
quantum mechanics 152–3, 173–8
quantum paradox, Schrödinger's cat 174
quantum tunneling 178
quarks 245–8
Queen Sonduk (7th-century Korea) 88

Rabi, I.I. (physicist), on crisis of physics 196
radar 219–20
radio, invention of 128–9
Radio K.A.O.S. (Roger Waters) 127
radio waves 127–33
RCA, television invention and 145
Reagan administration, "Star Wars" strategic defense initiative 218
Reddi-Wip 203, 205–6
reflecting telescope 29–33
Reflections on the Motive Power of Fire (Sadi Carnot) 84
refraction 19–20
refractor telescope 28, 29
Reichmann, Georg Wilhelm (lightning) 45
Reines, Frederick (neutrino) 268
Rhäticus (Rheticus), Georg Joachim, publishing *De revolutionibus* 12–13
Riche, Gaspard, Baron de Prony (French mathematician) 69
robots 212, 252–9

Rocket into Interplanetary Space The (Hermann Oberth) 169
rockets 166–71
Roddenberry, Gene (Star Trek) 238
Röntgen, Wilhelm (X-ray) 135–8
roller coasters 102–9
R.U.R. (Karel Čapek) 259

Sailor, Michael ("smart dust" creator) 225–6
Sandia National Laboratories, Z pinch 65
Sanskrit poems, 12th-century, using phi 3
Santos-Dumont, Alberto (dirigible) 78–9
Sayers, Dorothy (*Gaudy Night*) 272–3, 278
Scenic Railway 105
Schawlow, Arthur (stimulated emission) 220–1
Schön, Jan Hendrik (conductivity fraud) 275–8
Schrödinger, Erwin (Austrian physicist): paradox of Schrödinger's cat 174; particle-wave duality studies 177–9
Schrödinger Cat, The ballet 174
Schrödinger's Cat Trilogy (Robert Anton Wilson) 174
Schwinger, Julian (QED) 196, 201–2
science, women in 87–93
Science of Harry Potter, The (Roger Highsmith) 187
Scott, Ridley (*Blade Runner*) 252, 259
Search, The (C. P. Snow) 272
Searle, John (Chinese room parable) 254
Secret Knowledge (David Hockney) 50
Seife, Charles (*Zero: The Biography of a Dangerous Idea*) 153
sensitive dependence on initial conditions 233
shaving foam, development of 205–6
Shelter Island conference 195–6
Shelton, Ian (supernova 1987A discovery) 271
Shenandoah (dirigible) 78

Sidereus nuncius (*Starry Messenger*) (Galileo Gallilei) 29

Smalley, Richard (nanotechnology) 229–30

Smyth, AEC Commissioner Henry DeWolf 194

SNO (Sudbury Neutrino Observatory) 269–70

Snow, C. P. (*The Search*) 272

Sohn, Lydia (Schön fraud discovery) 275–6

solar neutrinos 266–70

Somerville, Mary (British mathematician/scientist) 91–2

sound waves 95, 98

special relativity 157–62

spectroscopic projection 119–20; *see also* camera obscura

speed of light: special relativity and 157–8; time travel and 162–5

spherical aberration 21

spiral nebulae, discovery of 31

spyglass, *see* telescope

St. Jerome in the Wilderness (Leonardo da Vinci) 5–6

"standard model" 246–8, 281

Star Trek 218, 223, 238, 242

Star Wars series 218, 221

"Star Wars" strategic defense initiative 218, 242

static electricity 44–6

Steadman, Philip (*Vermeer's Camera*) 53–4

Stepford Wives, The 253

Sterling, Bruce (*Difference Engine, The*) 71

Stevenson, Robert Louis (*Dr. Jekyll and Mr. Hyde*) 148, 156

stimulated emission, lasers and 218–20

Stork, David (Consulting Professor of Electrical Engineering) 57–8

string theory 279, 282–5

strong nuclear force 280–1

subatomic particles, *see* specific particles: neutrinos 266–71; overview 246–7; tracing 247–50

Subtle Knife, The (Philip Pullman) 260–1

Sudbury Neutrino Observatory (SNO) 269–70

Super-Kamiokande neutrino detector 271

superheroes, dual personality of 148–9

supernovas, neutrinos and 271

Survivor 233

Susskind, Leonard, string theory 282–3

sympathetic resonance 98

Szilard, Leo (atomic energy) 189–90

Talbot, William Henry Fox (calotype) 122

Taylor, Richard (fractal patterns in art) 235

telegraph, development of 95–7

Telegraph, radio waves and 129–31

telephone 97–101

telescope 26–33

television 144–7

Teller, Edward (atomic bomb development) 193–4

Terminator 2: Judgment Day (James Cameron) 192

Tesla, Nikola: AC power development 111–14; far thinking of 130–3; radio invention 128–9, 131–3

Tevatron atom smasher 248

thermodynamics: forces of, rockets and 166; foundation of 85–6; laws of 82–5; "Maxwell's demon" and 85–6; Planck's experiments with 151–5; quantum mechanics and 152–3

"thinking machine" 68–9

Thompson, La Marcus (roller coasters) 104–5

Thompson, Phil (fractal music) 236

Thompson, William (Lord Kelvin), foam studies 208

Thomson, Elihu (X-rays) 138

Thomson, Joseph John (J. J.) (cathode ray) 143–4

time dilation 160–3

time travel 157–8, 162–5

To Say Nothing of the Dog (Connie Willis) 164

Tolkien, J. R. R. (*The Fellowship of the Ring*) 225

Tomonaga, Sin-Itiro (Shinichiro) (QED) 196

top quarks 250–1

Townes, Charles (stimulated emission) 219–21

Trinity test (atomic bomb) 191–2

"Turk" chess player 254–5

Universal Foam (Sidney Perkowitz) 205–6

universe: basic building blocks of 246–7; expansion of 262–5; theories about 243; uranium 235, atomic bomb development and 190

Uranus, discovery of 30

van Eyck, Jan (artist), potential use of optical aids 56–7

van Leeuwenhoek, Antoni (microscopic) 24–5

van Musschenbroek, Pieter (Leyden jar) 45

Velcro 210–11, 213–14

Veneziano, Gabriele, string theory 282

Venus de Milo, phi and 3

Vermeer, Jan (artist) 52–4

Vermeer's Camera (Philip Steadman) 52–4

Verne, Jules (*From the Earth to the Moon*) 166, 169

Vezzosi, Alessandro (director of Museo Ideale) 7–8

Virtuoso, The (Thomas Shadwell) 23–4

Vitruvian Man (Leonardo da Vinci) 5–6

Vitruvius (Roman architect and engineer) 212

Volta, Alessandro (invention of battery) 48–9

von Braun, Werner (rocket) 170–1

von Kempelen, Baron Wolfgang (German) 212, 254

von Zeppelin, Count Ferdinand (dirigible) 78

Voss, Richard (fractal patterns in art) 235–6

War of the Worlds, The (H. G. Wells) 217

Waters, Roger (*Radio K.A.O.S.*) 127

waterwheel, perpetual motion efforts 82–4

waves, quantum mechanics and 176–80

Weaire, Dennis (foam) 208

weak nuclear force 280–1

weather forecasting, unpredictability of 231–2

Wedgwood, Thomas (photography) 120

weight, gravity and 35–6

Welles, Orson (*The War of the Worlds* radio broadcast) 217

Wells, H. G. (*The War of the Worlds*) 217

Westinghouse, George (AC power development) 112

Wheatstone, Charles (telegraph development) 96

Wieman, Carl (JILA physicist) 222–3

Wilder, Thornton (*Our Town*) 268

Willis, Connie: "At the Rialto" 173–5, 179–80; *The Doomsday Book* 164; *To Say Nothing of the Dog* 164

Wilson, Robert Anton (*Schrödinger's Cat Trilogy*) 174

Wilson, Robert (cosmic microwave background radiation) 243–4

Winterson, Jeanette (*Gut Symmetries*) 174–5, 189, 279, 293

Wise, James (fractal patterns) 237

witchcraft, educated women and 89

Witten, Ed (string theorist) 283

Wollaston, William (camera lucida) 54–5

women, in mathematics/science 87–93

World War II, atomic bomb development and 188–90

Wright, Orville (flight) 79, 212

Wright, Wilbur (flight) 79, 212

X: The Man with X-ray Eyes 134, 138

xerography 185

Xerox Corporation 186
X-Files, The 42, 252
X-Men comic books 59
X-rays 134–40

Young, Thomas (light) 150–1

Z pinch 65
Zero: The Biography of a Dangerous Idea
    (Charles Seife) 153
zoetrope 123
zoopraxiscope 123–5
Zworykin, Vladimir (television) 145